Magnetometers - Fundamentals and Applications of Magnetism

Edited by Sergio Curilef

Published in London, United Kingdom

IntechOpen

Supporting open minds since 2005

Magnetometers - Fundamentals and Applications of Magnetism
http://dx.doi.org/10.5772/intechopen.75335
Edited by Sergio Curilef

Contributors
Hsini Mohamed, Hector Lopez Loera, Myeongwon Lee, Jungbae Yoon, Donghun Lee, Navin Khaneja, Sergio Curilef, Sadok Zemni

First published in London, United Kingdom, 2020 by IntechOpen
IntechOpen is the global imprint of INTECHOPEN LIMITED, registered in England and Wales, registration number: 11086078, 5 Princes Gate Court, London, SW7 2QJ, United Kingdom
Printed in Croatia

British Library Cataloguing-in-Publication Data
A catalogue record for this book is available from the British Library

Additional hard and PDF copies can be obtained from orders@intechopen.com

Magnetometers - Fundamentals and Applications of Magnetism
Edited by Sergio Curilef
p. cm.
Print ISBN 978-1-83969-095-2
Online ISBN 978-1-83969-096-9
eBook (PDF) ISBN 978-1-83969-097-6

For EU product safety concerns:
IN TECH d.o.o., Prolaz Marije Krucifikse Kozulić 3, 51000 Rijeka, Croatia, info@intechopen.com or visit our website at intechopen.com.

Meet the editor

Sergio Curilef has been a full professor in the Physics Department, Universidad Católica del Norte, Antofagasta, Chile, since 1998. He obtained his PhD in Physics from the Brazilian Center for Research in Physics (CBPF)-RJ, Brazil, in July 1997. His main research areas are statistical physics and quantum mechanics of systems with long-range interparticle interactions, particles in the magnetic field, and nonlinear diffusion. He has recently developed research on complex systems and applications to science and engineering. He is the principal investigator in several research projects, and he did a sabbatical year at the University of Pretoria, South Africa, in 2007. He has also created the undergraduate and postgraduate programs in Physics at his university where he is currently director of the Physics Department.

Contents

Preface

Knowledge of magnetism spans from Ancient Greece to present day. Current theories of magnetism bring to light new applications, new features, and new models.

Magnetism is the source of two phenomena in physics: electric current and spin magnetic moments of elementary particles. As such, much attention is paid to low-dimensional structures. As the space dimension of a physical system decreases, magnetic ordering tends to vanish as fluctuations become quite significant. As known, there is no spontaneous magnetization in systems in one dimension at any nonzero temperature; for instance, the isotropic spin-s Heisenberg model, hard-core, and any system with finite range interactions. Nevertheless, the mean-field approximation constitutes an example of the state of magnetic ordering of a chain of spins with long-range microscopic interactions giving anomalous ferromagnetism cases in one dimension. Therefore, an accurate description of magnetic ordering phases illustrates concepts about the critical behavior and phase change and possible applications of new magnetic devices.

Following the Introductory chapter, four chapters cover topics related to recent advances in the modeling and application of magnetometers.

M. Hsini, S. Zemni, In Chapter 2 "Modeling the Magnetocaloric Effect of $Nd_{0.67}Ba_{0.33}Mn_{0.98}Fe_{0.02}O_3$ by the Mean Field Theory," M. Hsini and S. Zemni combine mean-field theory and the Bean–Rodbell model to justify the magnetocaloric effect (MCE) in a sample. They derive expressions to rating the magnetic entropy change under various magnetic fields and compare theoretical to experimental curves.

H. López Loera, In Chapter 3 "The Magnetometry—A Primary Tool in the Prospection of Underground Water," H. López Loera presents a geophysical methodology that shows the potential of combining natural and induced methods to locate confined aquifers in zones with a high probability of locating groundwater in the Mexican Mesa Central.

In Chapter 4 "Atomic Scale Magnetic Sensing and Imaging Based on Diamond NV Centers," M. Lee, J. Yoon, and D. Lee review the basic sensing mechanisms of the nitrogen-vacancy (NV) center and introduce imaging applications based on scanning magnetometry and wide field-of-view optics.

Finally, in Chapter 5 "SQUID Magnetometers, Josephson Junctions, Confinement and BCS Theory of Superconductivity," N. Khaneja discusses some theoretical aspects of a SQUID magnetometer, as its sensitivity makes it possible to apply to biomagnetism, materials science, metrology, astronomy, and geophysics.

Sergio Curilef
Universidad Católica del Norte,
Antofagasta, Chile

Section 1

Introduction

Chapter 1

Introductory Chapter: Statistical and Theoretical Considerations on Magnetism in Many-Body Systems

Sergio Curilef

1. Long-range interactions

The description of systems with long-range interactions is relevant to statistical physics because we find appealing properties that deserve to be studied in detail. In this line, some variations of the Ising model that involves not only first neighbors, but also distant neighbors are employed; for instance, the Hamiltonian Mean Field (HMF), and the recently introduced dipole-type Hamiltonian Mean Field (d-HMF) model. We emphasize that the Ising model is a recurrent tool to study magnetic properties and the statistical behaviors of many-body systems in the broadest context [1–5].

In concern about long-range interactions, there are various challenges to face; these are related to the dynamics, size of the systems, and the theoretical framework to explain the behavior of systems, among others. In this type of system, we have typical consequences such as the loss of additivity and extensivity. The loss of additivity takes place for ensembles of interacting particles that cannot trivially separate into independent subsystems, which is explained by the presence of underlying interactions or correlation effects, whose characteristic lengths are comparable or more significant than the system linear size [6–8]. Additionally, the loss of extensivity frequently accompanies the loss of additivity [5, 9].

If we have a system composed of N spins that interact one to each other as a power law of inter-particle distance, the total energy of the system is E. The system divided into two subsystems, 1 and 2, both composed of $N/2$ particles with energies E_1 and E_2, respectively. If the spins in subsystem 1 are up, and those in the subsystem 2 are down, the energies satisfy $E_1 = E_2$. Since the sum of energies, $E_1 + E_2$ is not equal to the total energy E; the system is nonadditive. In general, $E < E_1 + E_2$. Besides, the extensivity ($E/N \rightarrow$constant) is another fundamental property to consider in this kind of analysis. The recent literature shows a way to recover this property through Kac's prescription that has a standard thermodynamic structure because it preserves the Euler and Gibbs-Duhem relations. Therefore, this procedure allows us to recover the extensivity, while the loss of additivity remains because of the long-range interactions [5, 9].

At equilibrium, an analytical procedure leads to solving the problem for obtaining the magnetization, the inverse temperature, the specific heat, in the canonical ensemble. Also, it is possible to get the microcanonical entropy. The caloric curve, ascertained using both, exactly coincides. At this stage, we emphasize that the solution to this system with long-range interactions becomes analytical, which solutions are not abundant in statistical physics. Nevertheless, out of equilibrium, we have data from simulations obtained from carrying out molecular dynamics.

IntechOpen

The evolution of simulations permits us to observe several properties that are not possible to identify with any theoretical description. The interpretation comes from perceiving regularities in the dynamics of systems as HMF and d-HMF models.

2. Equilibrium

As generally noticed, at low energy, a phase of a single cluster of particles appears; at high energy, a homogeneous phase emerges.

For theoretical and numerical modeling, we consider a system of N identical, coupled, dipole-type particles with a mass equal to 1. The dynamics evolves in a periodic cell described by a one-dimensional, dipole-type HMF model (d-HMF) [9, 10] given by

$$H=\frac{1}{2}\sum_{i=1}^{N}p_i^2+\frac{\lambda}{2N}\sum_{i\neq j}^{N}\left[\cos\left(\theta_i-\theta_j\right)-3\cos\theta_i\cos\theta_j-\Delta_{i,j}\right], \tag{1}$$

where the variable p_i is the momentum of the particle i, and θ_i is its corresponding angle of orientation (integer $i\in[1,N]$ for the system size N). The parameter λ stands for the coupling and $\Delta_{i,j}$ suitably denotes the zero of the potential.

The equations of motion are derived by the following set:

$$p_i=\frac{-\varepsilon}{2}\left(2M_x\sin\theta_i-M_y\cos\theta_i\right), \tag{2}$$

where the components of the magnetization vector are defined as

$$\left(M_x,M_y\right)=\frac{1}{N}\sum_{i\neq j}^{N}(\cos\theta_i,\sin\theta_i). \tag{3}$$

After deriving thermodynamic properties in both, the canonical and microcanonical ensembles, the caloric curves come from the following coincident result:

$$2\beta=\frac{1}{\left(\varepsilon+\lambda\left(m^2+\Delta/2\right)\right)}, \tag{4}$$

combined with the extremal solution

$$m=\frac{I_1(2\beta\lambda m)}{I_0(2\beta\lambda m)}, \tag{5}$$

where $I_k(\ldots)$ is the kth-order modified Bessel function, β is the inverse of temperature, ε is the energy per particle; m is the magnetization per particle. We analytically obtain the parameters that define the critical point, such as the critical inverse temperature β_c = 1, critical internal energy ε_c = 3/2, and critical magnetization m_c = 0. The specific heat diverges with the temperature $1/\beta\rightarrow1^-$ and remains in a constant value for $1/\beta>1$, which corresponds to an ideal gas in one dimension. Therefore, systems with magnetic behavior are recently studied in [11, 12], giving an approach to the understanding of magnetic materials.

As said before, this system denotes an analytically solvable many-body problem. This type of solutions is not common in physics and particularly in statistical mechanics, especially regarding long-range interactions [9]. Baxter gives a comprehensive overview in his book [13]. The spherical model is an additional solvable problem that includes long-range interactions.

3. Quasi-stationary states

The statistical mechanics thoroughly explains the equilibrium, which is the statistical state reached in the long-lasting evolution. Nevertheless, in intermediate intervals of the evolution time, the dynamics is abnormal, and the description is not complete. Therefore, there are at least two regimes in the evolution of systems with long-range interactions: the equilibrium and the quasi-stationary states (QSS). In the first case, we have the Boltzmann-Gibbs statistics to describe the state of the systems entirely. In the second, it is an open problem that possesses several theoretical attempts susceptible to improve them.

Generally speaking, when the modeling neglects all of the spatial structure of the system but considers long-range interactions, mean-field approximations should be good alternatives to consider. The d-HMF model is curious because it neglects the structure of systems and space interactions, and only keeps the part related to orientations, but shows axiomatic and pertinent properties.

To identify the QSS, we search for intervals where the thermodynamic values keep constant. Therefore, **Figure 1(a)** illustrates two plateaus in the behavior of magnetization per particle as a function of time, before equilibrium. A complementary perspective of the first QSS given in [10] considers the shape of the mean kinetic energy per particle, aiding us to obtain the power-law duration depicted in **Figure 1(b)** that characterizes the nature of QSS. States defined by the plateaus of mean kinetic energy per particle are lower than the canonical temperature, whose values only coincide in equilibrium [10].

Finally, this system constitutes an example of a non-symmetric HMF that shows a phase transition, the appearance of a spontaneous magnetic ordering. The symmetry is compared to the HMF model that remains invariant under a typical rotational transformation, which is not valid for the d-HMF model, but presents a second-order phase transition; therefore, a possible application found in the literature is the phase transition of non-symmetric spin glasses. Theoretical background for systems out of equilibrium is not unique and fundamental questions related

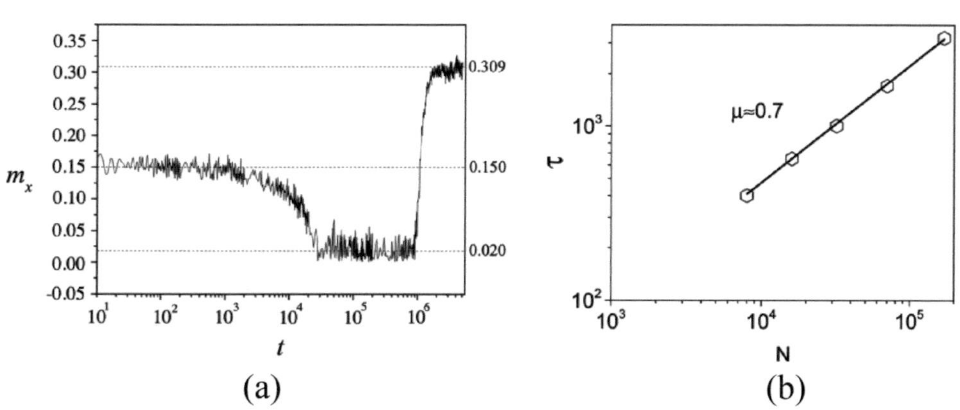

Figure 1.
The magnetization per particle is depicted as a function of the time in (a). In (b), the relaxation time τ, of the first QSS that goes to the second QSS, is depicted as a function of 1/N in log–log scale, whose power law of the duration of the first QSS in terms of the system size is obtained from the behavior of the kinetic energy.

to this issue are still open; a possible description based on a family of solutions of Vlasov equation is in progress [14].

The understanding of these thermodynamic properties can be fundamental in future demands as the manufacturing of magnetic instruments, sensors, and magnetometers.

Author details

Sergio Curilef
Departamento de Física, Facultad de Ciencias, Universidad Católica del Norte, Antofagasta, Chile

*Address all correspondence to: scurilef@ucn.cl

References

[1] Tatekawa T, Bouchet F, Dauxois T, Ruffo S. Thermodynamics of the self-gravitating ring model. Physical Review E. 2005;**71**:056111

[2] Campa A, Dauxois T, Ruffo S. Statistical mechanics and dynamics of solvable models with long-range interactions. Physics Reports. 2009;**480**:57-159

[3] Pluchino A, Latora V, Rapisarda A. Dynamics and thermodynamics of a model with long-range interactions. Continuum Mechanics and Thermodynamics. 2004;**16**:245-255

[4] Atenas B, Curilef S. Dynamics of the d-HMF model: Sensitive dependence on size and initial conditions. Journal of Physics Conference Series. 2018;**1043**:1-7

[5] del Pino LA, Troncoso P, Curilef S. Thermodynamics from a scaling Hamiltonian. Physical Review B. 2007;**76**:172402

[6] Dauxois T, Ruffo S, Arimondo E, Wilkens M, editors. Dynamics and thermodynamics in systems with long-range interactions. In: Lecture Notes in Physics. Vol. 602. Berlin: Springer; 2002

[7] Levin Y, Pakter R, Rizzato FB, Teles TN, da C Benetti FP. Nonequilibrium statistical mechanics of systems with long-range interactions. Physics Reports. 2014;**535**:1-60

[8] Curilef S, del Pino LA, Orellana P. Ferromagnetism in one dimension: Critical temperature. Physical Review B. 2005;**72**:224410

[9] Atenas B, Curilef S. A solvable problem in statistical mechanics: The dipole-type Hamiltonian mean field model. Annals of Physics. 2019;**409**:167926

[10] Atenas B, Curilef S. Dynamics and thermodynamics of systems with long-range dipole-type interactions. Physical Review E. 2017;**95**:022110

[11] Olivares F, Pennini F, Curilef S. Thermodynamics in a complete description of Landau diamagnetism. Physical Review B. 2010;**81**:041134

[12] Atenas B, Curilef S. Classical states of an electric dipole in an external magnetic field: Complete solution for the center of mass and trapped states. Annals of Physics. 2014;**350**:605-614

[13] Baxter RJ. Exactly Solved Models in Statistical Mechanics. Printed in Great Britain by St Edmundsbury Press Limited Bury St Edmunds. Suffolk: Academic Press; 1982

[14] Atenas B, Curilef S. A statistical description for the quasi-stationary-states of the dipole-type Hamiltonian mean field model based on a family of Vlasov solutions. Preprint submitted for publication; 2020

Section 2

Modeling

Chapter 2

Modeling the Magnetocaloric Effect of $Nd_{0.67}Ba_{0.33}Mn_{0.98}Fe_{0.02}O_3$ by the Mean Field Theory

Mohamed Hsini and Sadok Zemni

Abstract

In this paper, we have exploited the mean field theory combined with the Bean-Rodbell model to justify the magnetocaloric effect (MCE) in $Nd_{0.67}Ba_{0.33}Mn_{0.98}Fe_{0.02}O_3$ sample. The simulation of some magnetic properties has been investigated. Modeling magnetization curves have been successfully achieved using this model. The second-order ferromagnetic-paramagnetic (FM-PM) phase transition of our system has been verified through the value of the parameter which controls the transition nature in the Bean-Rodbell model. Theoretical and experimental expressions, which have rated the magnetic entropy change ($-\Delta S_M$) under various magnetic fields, have been derived. Theoretical ($-\Delta S_M$) curves have been compared to the experimental ones.

Keywords: mean field theory, Bean-Rodbell, magnetocaloric effect, magnetization, magnetic entropy change

1. Introduction

In recent years, magnetic materials exhibiting high magnetocaloric effect (MCE) have been extensively studied experimentally and theoretically because of their intensive necessity in magnetic refrigeration (MR) [1–3]. This recent cooling technology, which is expected to replace traditional expansion/compression gas refrigeration technology, has many particular interests because of its significant economic benefits [4–6]. The magnetic entropy change ($-\Delta S_M$) is interestingly important for rating the refrigerant properties [7, 8]. Thus, numerous materials exhibiting high MCE have been widely developed in the last decades for exploitation as promising materials in MR technology [9–14].

The mean field model [15, 16] is an efficient tool in the study of magnetic materials [17]. Currently, Amaral et al. have signaled a scaling method based on this model [18].

According to the work of Amaral et al., we have reported, in this paper, our studies on the magnetocaloric properties of the $Nd_{0.67}Ba_{0.33}Mn_{0.98}Fe_{0.02}O_3$ sample which exhibits a second-order ferromagnetic-paramagnetic (FM-PM) phase transition [19], by scaling the experimental magnetization. We have showed, in this work, how the mean field theory may adequately model the magnetic and the magnetocaloric properties of this magnetic system, which may be applicant in MR technology.

The scaling method based on the mean field model leads us to estimate directly the saturation magnetization (M_0), the exchange parameter (λ), the total angular momentum (J), and the Lande factor (g) of our sample. These parameters are necessary for simulating magnetization isotherms, M (H, T), which are used for the calculation of the magnetic entropy change ($-\Delta S_M$) of $Nd_{0.67}Ba_{0.33}Mn_{0.98}Fe_{0.02}O_3$ material. In addition, the second-order phase transition FM-PM of this sample is confirmed by analyzing the Bean-Rodbell model [20, 21].

2. Theoretical and experimental study

2.1 Brief overview of the experimental study

In this section, we have summarized the primary results of the structural and magnetic analysis of the manganite sample $Nd_{0.67}Ba_{0.33}Mn_{0.98}Fe_{0.02}O_3$ reported in our precedent work [19].

This compound has been prepared by the solid-state ceramic method at 1400°C in a polycrystalline powder form. Rietveld structural analysis has showed a good crystallization of the sample which presents a pseudo-cubic structure of orthorhombic Imma distortion, with unit cell parameters a = 0.54917 (1), b = 0.77602 (1), c = 0.551955 (4) nm, and unit cell volume V = 0.235228 (3) nm^3. Scanning electron microscope (SEM) analysis has indicated that the sample presented a homogeneous morphology which consists of well-formed crystal grains. The SEM analysis coupled with the EDX has confirmed that the chemical composition of the sample is close to that nominal reported by the above chemical formula (19).

The evolution of the magnetization as a function of temperature, under a 0.05 T magnetic applied field in FC and ZFC modes, is depicted in **Figure 1**. This figure shows a FM-PM transition at a Curie temperature which has been estimated in the inset by determining the minimum value of the derivative magnetization versus temperature in ZFC mode at 0.05 T applied field ($T_C = 131$ K). However, **Figure 1** shows a non-negligible monotonic decrease of the magnetization between 10 and 100 K. This indicates a canted spin state between the Nd^{3+} and the (Mn^{3+}, Mn^{4+}) spin sub-lattices, with canted angle, θ, assumed to be between 0° (ferromagnetic coupling) and 180°(antiferromagnetic coupling).

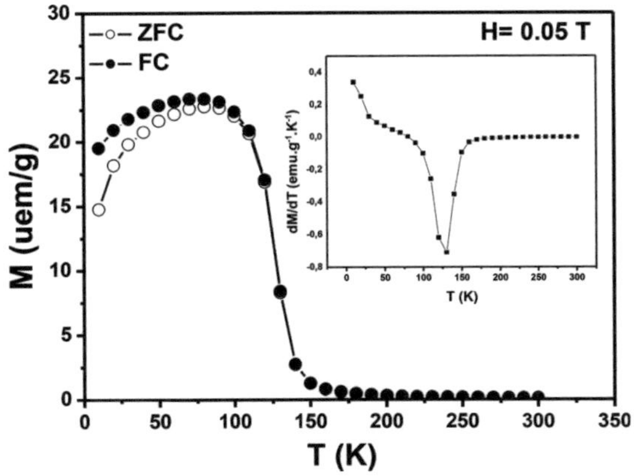

Figure 1.
M *versus* T *in 0.05 T applied magnetic field for the* $Nd_{0.67}Ba_{0.33}Mn_{0.98}Fe_{0.02}O_3$ *versus* T *in FC and ZFC modes. The inset is* $\frac{dM}{dT}$ *versus* T *for ZFC mode.*

Figure 2 shows the variation of the magnetization as a function of the varied magnetic field up to 10 T, at very low temperature (10 K), for the undoped compound $Nd_{0.67}Ba_{0.33}MnO_3$ and for the doped compound $Nd_{0.67}Ba_{0.33}Mn_{0.98}Fe_{0.02}O_3$.

It is apparent in this figure that in spite of the intense magnetic applied field (10 T), the magnetization does not attain saturation. This is due to the presence of the magnetic moments of Nd^{3+} ([Xe] $4f^3$) which have three electrons in the 4f orbital. Effectively, a comparison between magnetization of the two compounds $Nd_{0.67}Ba_{0.33}MnO_3$ [19] and $La_{0.67}Ba_{0.33}MnO_3$ [22] are depicted in **Figure 3**. This figure shows obviously that the lanthanum compound rapidly reaches saturation even under low applied magnetic field. This is because of the non-contribution of the La^{3+} ion ([Xe]) in magnetism which has no electrons in 4f orbital. **Figure 2** also indicates that a 2% iron doping proportion in $Nd_{0.67}Ba_{0.33}Mn_{0.98}Fe_{0.02}O_3$ decreases the magnetization by $0.12\mu_B$ ($3.94\ \mu_B$ for $Nd_{0.67}Ba_{0.33}MnO_3$, whereas $Nd_{0.67}Ba_{0.33}Mn_{0.98}Fe_{0.02}O_3$ presents $3.82\ \mu_B$) under a 10 T applied magnetic field of, in a good agreement with an antiferromagnetic coupling between Mn^{3+} and Fe^{3+} spin sub-lattices as demonstrated by the Mössbauer spectroscopy studies [23, 24]. As knowing, the orbital momentum is quenched by the crystal field in the octahedral site of manganite for transition elements, so only the spin of Fe^{3+} ([Ar]$3d^5$) contributes to

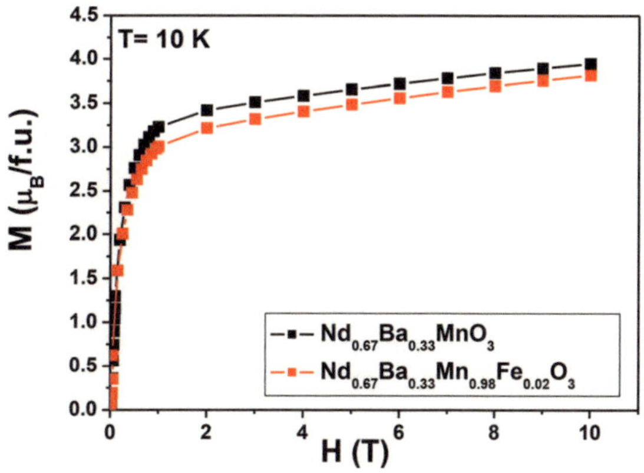

Figure 2.
Comparison of M *versus* H *at* T = 10 K *for* $Nd_{0.67}Ba_{0.33}MnO_3$ *and* $Nd_{0.67}Ba_{0.33}Mn_{0.98}Fe_{0.02}O_3$ *samples.*

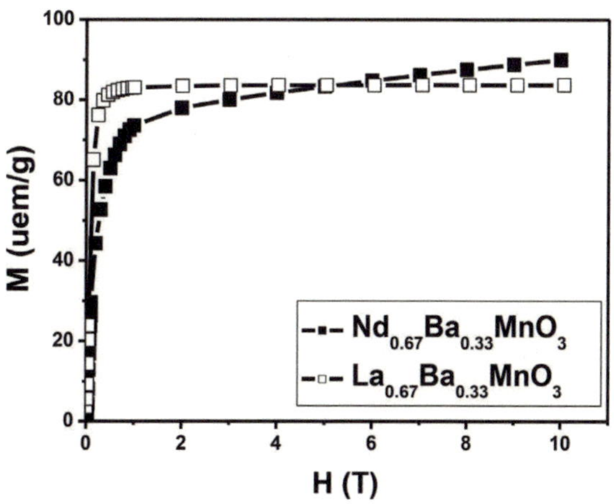

Figure 3.
Comparison of M *versus* H *at* T = 10 K $La_{0.67}Ba_{0.33}MnO_3$ *and* $Nd_{0.67}Ba_{0.33}MnO_3$ *for samples.*

the magnetization; therefore, we have found that the experimental value ($0.12\mu_B$) is very near to that calculated ($M_{Fe^{3+}} = gS\mu_B = 0.02 \times 2 \times 5/2\ \mu_B = 0.1\mu_B$).

2.2 Theoretical calculation

The magnetic moments of a ferromagnetic material, made under external magnetic field (*H*), tend to align in the *H* direction. The increase of parallel magnetic moments then leads to rising magnetization. Magnetization values could be rated by the Weiss mean field theory [15, 16, 18].

In fact, Weiss has enunciated that in a ferromagnetic, an exchange interaction between magnetic moments could be created, at least in a magnetic domain, where the magnetic moments could be ordered in a same direction. This interaction may be considered as an average over all interactions between a given magnetic moment and the other N magnetic moments of the Weiss domain. This internal interaction contributes to an exchange field or a Weiss mean field:

$$\vec{H}_W = \vec{H}_{exch} = \lambda \vec{M} \tag{1}$$

where λ is the exchange parameter and M is the magnetization of the ferromagnet, given by

$$M = M_0 B_J(x) \tag{2}$$

where

$$M_0 = NJg\mu_B \tag{3}$$

is the saturation magnetization,

$$B_J(x) = \frac{2J+1}{2J}\coth\left(\frac{2J+1}{2J}x\right) - \frac{1}{2J}\coth\left(\frac{x}{2J}\right) \tag{4}$$

is the Brillouin function, and

$$x = \frac{Jg\mu_B}{k_B}\left(\frac{H + H_{exch}}{T}\right) \tag{5}$$

where k_B is the Boltzmann constant, μ_B is the Bohr magnetron, N is the number of spins, and T is the temperature.

Eq. (2) can be written as a function of $\left(\frac{H+H_{exch}}{T}\right)$ as follows:

$$\begin{aligned} M(H,T) = f\left(\frac{H+H_{exch}}{T}\right) &= M_0\left[\frac{2J+1}{2J}\coth\left(\frac{2J+1}{2J}\frac{Jg\mu_B}{k_B}\left(\frac{H+H_{exch}}{T}\right)\right)\right. \\ &\left. -\frac{1}{2J}\coth\left(\frac{1}{2J}\frac{Jg\mu_B}{k_B}\left(\frac{H+H_{exch}}{T}\right)\right)\right] \end{aligned} \tag{6}$$

Applying the reciprocal function f^{-1} of f, we can obtain the relations:

$$\frac{H}{T} = f^{-1}(M) - \frac{H_{exch}}{T};\ H = Tf^{-1}(M) - \lambda M. \tag{7}$$

The magnetic entropy change can be expressed by the Maxwell relations [6, 25]:

$$\left(\frac{\partial S}{\partial H}\right)_T = \left(\frac{\partial M}{\partial T}\right)_H \tag{8a}$$

$$\left(\frac{\partial S}{\partial M}\right)_T = -\left(\frac{\partial H}{\partial T}\right)_M \tag{8b}$$

Using experimental isotherm magnetization data, measured at discreet values of both applied magnetic field and temperatures, and Eq. (8a), the magnetic entropy change can be approximated as

$$\Delta S_M(T, \Delta H) = \int_{H_1}^{H_2} \left(\frac{\partial M}{\partial T}\right)_H dH \approx \sum_n \frac{(M_{n+1} - M_n)_H}{T_{n+1} - T_n} \Delta H_n \tag{9a}$$

Eqs. (7) and (8b) allow us to determine the theoretical estimation of the magnetic entropy change:

$$\Delta S_M(T)_{H_1 \to H_2} = -\int_{M|_{H_1}}^{M|_{H_2}} \left(f^{-1}(M) - \left(\frac{\partial H_{exch}}{\partial T}\right)_M\right) dM \tag{9b}$$

To study the nature of the magnetic transition, we have called the Bean-Rodbell model to our magnetization data. As reported earlier [26–28], system exhibiting first- or second-order phase transitions have been interpreted using this model [29]. It considers that exchange interactions adequately depend on the interatomic distances; the Curie temperature T_C is expressed as follows:

$$T_C = T_0(1 + \beta\omega) \tag{10}$$

where $\omega = \frac{v - v_0}{v_0}$, v is the volume, v_0 presents the volume with no exchange interaction, and T_0 is the transition temperature if magnetic interactions are taking into account with no magneto-volume effects. β is the slope of the critical temperature curve on volume. The Gibbs free energy, for a ferromagnetic system, is given in Ref. [30] with the compressibility K, the magnetic entropy S, and the reduced magnetization $\sigma(x) = B_J(x)$ as

$$G = -\frac{3}{2}\left(\frac{J}{J+1}\right) N k_B T_C \sigma^2 - HgJ\mu_B N\sigma + \frac{1}{2K}\omega^2 - TS \tag{11}$$

The above free energy minimizes $\left(\frac{dG}{d\omega} = 0\right)$ at

$$\omega = \frac{3}{2}\left(\frac{J}{J+1}\right) N k_B K T_C \sigma^2 \tag{12}$$

Substituting Eq. (12) into Eq. (11) and minimizing G with respect to σ, according to the work of Zach et al. [29] and Tishin and Spichkin [6], we can obtain the magnetic state equation:

$$\sigma(Y) = B_J(Y) \tag{13}$$

with

$$Y = \frac{1}{T}\left[3T_0\left(\frac{J}{J+1}\right)\sigma + \frac{gJ\mu_B}{k_B}H + \frac{9}{5}\frac{(2J+1)^4 - 1}{[2(J+1)]^4} T_0 \eta \sigma^3\right] \tag{14}$$

where the parameter η checks the order of the magnetic phase transitions. For $\eta > 1$, the transition is assumed to be first order. For $\eta < 1$, the second-order magnetic phase transition takes place.

After combining Eq. (2) and Eq. (13), we have got two interesting equations, $M(x) = M_0B_J(x)$ (giving simulated M versus H) and $M(Y) = M_0B_J(Y)$ (giving simulated M versus T).

On the other hand, for weak values of x, the magnetization may be written as

$$M = M_0 g\mu_B \frac{H}{kT}\frac{J+1}{3} + \lambda M_0 g\mu_B \frac{1}{kT}\frac{J+1}{3} M = \chi H \tag{15}$$

The resolution of Eq. (15) gives easily the Curie-Weiss magnetic susceptibility:

$$\chi = \frac{NJ(J+1)g^2\frac{\mu_B^2}{3k_B}}{T - \lambda NJ(J+1)g^2\frac{\mu_B^2}{3k_B}} = \frac{C}{T - T_c} \tag{16}$$

where $T_c = \lambda C$ is the Curie temperature and C is the Weiss constant.

To determine accurately the exchange constant, we use the famous law of interaction between two magnetic atoms with spins $\vec{S_1}$ et $\vec{S_2}$ by the Hamiltonian [12]:

$$H = -2J\vec{S}_1\vec{S}_2 \tag{17}$$

given by Heisenberg, where **J** is the exchange constant.

If we consider an individual atom i with its magnetic moment $\vec{\mu}_i$ in a ferromagnetic system. This moment interacts with the external applied magnetic field $\vec{H}$ and with the exchange field $\vec{H}_{exch}$. The total interaction energy is given as

$$E_i = -\vec{\mu}_i\left(\vec{H} + \vec{H}_{exch}\right) \tag{18}$$

From the Heisenberg model's viewpoint, the energy E_i of a ferromagnetic system is the sum of interaction energy of a given moment, $\vec{\mu}$, with the external field and that with all near neighbors to atom i. Let us consider that each atom has z near neighbors which can interact with spin i with the same force, i.e., that exchange parameter has the same value for all z neighbors. Then, the energy E_i may be written as

$$E_i = -\vec{\mu}_i\vec{H} - \sum_{k=1}^{Z} 2J\vec{S}_i\vec{S}_k \tag{19}$$

where the index k runs over all z neighbors of the atom i.

It is practical to express the Heisenberg energy in Eq. 17 in terms of the atomic moments $\vec{\mu}$ rather than in terms of spins $\vec{S}$. It can be easily done if we consider that the relation between the spin and the atomic magnetic moment is $\vec{\mu} = -g\mu_B\vec{S}$. So, Eq. (19) would be rewritten as

$$E_i = -\vec{\mu}_i\left(\vec{H} + \frac{2J}{(g\mu_B)^2}\sum_{k=1}^{z}\vec{\mu}_k\right) \tag{20}$$

In fact, the two expressions of the energy E_i, in Eq. (18) and Eq. (20), are not similar. However, the sum term in Eq. (20) is not the same as $\vec{H}_{exch}$ because it is the interaction made by the individual spin i. But, $\vec{H}_{exch}$ in the spirit of the Weiss theory presents the average of all interaction terms in the total system. As a result, we should carry out such averaging over all N atoms:

$$\vec{H}_{exch} = \left\langle \frac{2J}{(g\mu_B)^2} \sum_{k=1}^{z} \vec{\mu}_k \right\rangle = \frac{2J}{(g\mu_B)^2} \frac{1}{N} \left\langle \sum_{i}^{N} \sum_{k=1}^{Z} \vec{\mu}_{i,k} \right\rangle \quad (21)$$

By summing on k, we could obtain

$$\vec{H}_{exch} = \frac{2J}{(g\mu_B)^2} \frac{1}{N} z \left\langle \sum_{i}^{N} \vec{\mu}_i \right\rangle = \frac{2J}{(g\mu_B)^2} \frac{1}{N} z \, \vec{M},$$

(then,)

$$E_i = -\vec{\mu}_i \left(\vec{H} + \frac{2Jz}{(g\mu_B)^2 N} \vec{M} \right) \quad (22)$$

Eq. (21) contains a term proportional to the magnetization, in perfect agreement with Weiss' postulate. We can write now

$$\vec{H}_{exch} = \frac{2Jz}{(g\mu_B)^2 N} \vec{M} \quad (23)$$

from which we immediately obtain the formula for Weiss' "effective field constant":

$$\lambda = \frac{2Jz}{(g\mu_B)^2 N} \quad (24)$$

Comparing this equation with Eq. (16), i.e., the phenomenological expression for λ, we obtain the solution for the critical temperature:

$$T_c = \lambda C = \frac{2Jz}{(g\mu_B)^2 N} \frac{NJ(J+1) g^2 {\mu_B}^2}{3k_B} \quad (25)$$

Therefore

$$J = \frac{3k_B T_c}{2zJ(J+1)} \quad (26)$$

2.3 Mean field theory application

We begin by the determination of J, g, λ, and M_0 parameters, which are crucial for magnetocaloric effect simulation of $Nd_{0.67}Ba_{0.33}Mn_{0.98}Fe_{0.02}O_3$.

- Total angular momentum (J) determination

To determine the total angular momentum (J), we must quantify the canted spin angle, θ, between $Nd^{3+}$and (Mn^{3+}, Mn^{4+}) spin sub-arrays, using the difference between magnetizations of $Nd_{0.67}Ba_{0.33}MnO_3$ [19] and $La_{0.67}Ba_{0.33}MnO_3$ [22] samples at 10 T ($0.33\mu_B$) as shown in **Figure 3**. By writing the contribution of Nd^{3+} magnetic moment network (spin-orbit coupling) under the form

$M_{Nd^{3+}} = 0,67\, J_{Nd^{3+}} g_{Nd^{3+}} \mu_B \cos\theta = 0.33\mu_B$, where $J_{Nd^{3+}} = 4.5$ and $g_{Nd^{3+}} = 0.727$ are, respectively, the values of angular momentum and gyromagnetic factor for free ion Nd^{3+} as indicated in Ref. [16]. Therefore, we deduce.

$$\cos\theta = \frac{0,33}{0.67J(Nd^{3+})\, g(Nd^{3+})} = \frac{0.33}{0.67 \times 4.5 \times 0.727} = 0.15, \text{ so, } \theta = 81.34^{\circ}.$$

Using the Hund's rule for 4f orbital less than half full and the values of L and S indicated in Ref. [16] for Nd^{3+}, we obtain the value of the angular momentum of Nd^{3+} ion incorporated in $Nd_{0.67}Ba_{0.33}Mn_{0.98}Fe_{0.02}O_3$ sample:

$$J(Nd^{3+}) = 0.67 \times |L - S \times \cos\theta| = 0.67 \times \left|6 - \frac{3}{2} \times 0.15\right| = 3.869$$

As a result, the total angular momentum for $Nd_{0.67}Ba_{0.33}Mn_{0.98}Fe_{0.02}O_3$ sample is

$$\begin{aligned} J &= J(Nd^{3+}) + S(Mn^{3+}) + S(Mn^{4+}) - S(Fe^{3+}) \\ &= 3.869 + 0.65 \times 2 + 0.33 \times 1.5 - 0.02 \times 2.5 = 5.614. \end{aligned}$$

- Gyromagnetic factor (g) determination:

$$g(Nd^{3+}) = 1 + \frac{J(J+1) + S\cos\theta(S\cos\theta + 1) - L(L+1)}{2J(J+1)} = 1 + \frac{5.754 \times 6.754 + 0.246 \times 1.246 - 6 \times 7}{2 \times 5.754 \times 6.754} = 0.96.$$

for all the sample $g = 0.67 \times 0.96 + 0.65 \times 2 + 0.33 \times 2 + 0.02 \times 2 = 2.6432$.

Figure 4 shows the evolution of M versus H at different T near T_C for the $Nd_{0.67}Ba_{0.33}Mn_{0.98}Fe_{0.02}O_3$ compound. The isothermal M (H, T) curves show a dependency between M and H at different T. Above T_C, a drastic decrease of $M(H,T)$ is observed with an almost linear behavior indicating a paramagnetic behavior. Below T_C, the curves show a nonlinear behavior with a sharp increase for

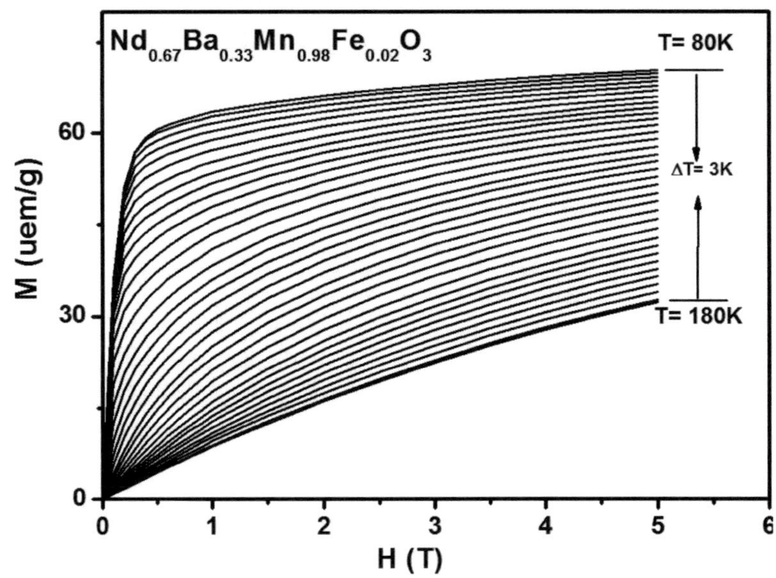

Figure 4.
Isotherm magnetization M as a function of magnetic field H, measured for different temperatures with a step of 3 K for $Nd_{0.67}Ba_{0.33}Mn_{0.98}Fe_{0.02}O_3$ sample.

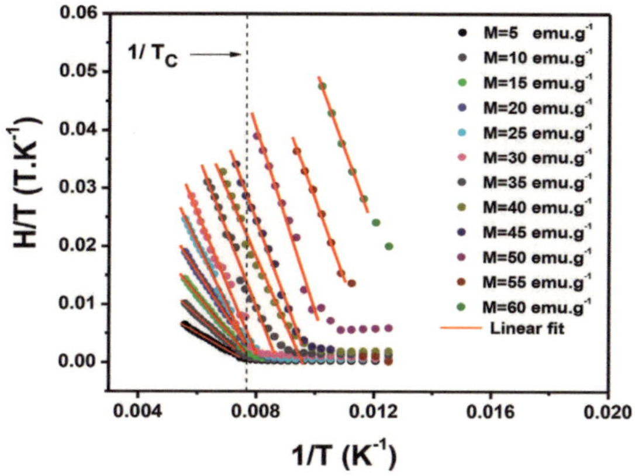

Figure 5.
$\frac{H}{T}$ versus $\frac{1}{T}$ curves with constant values of magnetization per curve for $Nd_{0.67}Ba_{0.33}Mn_{0.98}Fe_{0.02}O_3$ sample.

low field values and a tendency to saturation, as field increases, reflecting a ferromagnetic behavior. Using **Figure 4**, we could plot the evolution of $\frac{H}{T}$ versus $\frac{1}{T}$ taken at constant values of magnetization M (5 emu.g^{-1} step) from 180 to 80 K in **Figure 5**. A linear behavior of the isomagnetic curves, which are progressively shifted into higher $\frac{1}{T}$ values, could be observed. So, the linear relationship between $\frac{H}{T}$ and $\frac{1}{T}$ is preserved. To find the value of the parameter λ, it is necessary to study H_{exch} induced by magnetization change. Linear fits are then kept at each isomagnetics line. Using Eq. (6), the slope of each isomagnetics line could give the suitable H_{exch} value. In **Figure 6**, we have plotted H_{exch} vs. M for the $Nd_{0.67}Ba_{0.33}Mn_{0.98}Fe_{0.02}O_3$ compound. For all materials, in the PM or antiferromagnetic domain, we can always expand increasing H in powers of *M* or *M* in powers of *H*. In this approach, we will stop at the third order, and considering that the *M* is an odd function of *H* [24, 31], we can write

$$H_{exch} = \lambda_1 M + \lambda_3 M^3 \quad (27)$$

Then, these points in **Figure 6** (H_{exch} versus M) should be included for the fit by Eq. (27).

However, a very small dependence on M^3 ($\lambda_3 = -0.00006\ (\text{T.emu}^{-1}.\text{g})^3$) is noted for this second-order transition system. So, we can assume that $H_{exch} \approx \lambda M$,

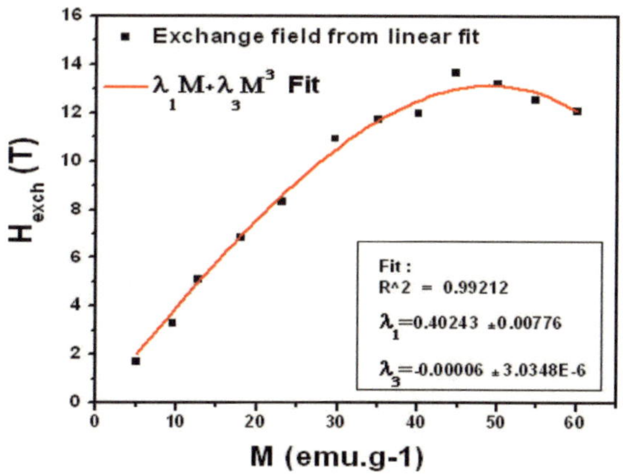

Figure 6.
Exchange field versus magnetization for $Nd_{0.67}Ba_{0.33}Mn_{0.98}Fe_{0.02}O_3$ sample, with the function $\lambda_1 M + \lambda_3 M^3$ fit.

with $\lambda = \lambda_1 = 0.40243$ T.emu^{-1}.g. Next, the building of the scaling plot M versus $\frac{H+H_{exch}}{T}$ is depicted in **Figure 7** with black symbols. It is clear from this figure that all these curves converge into one curve which can be adjusted by Eq. (6) using MATLAB software to determine M_0, J, and g. We have found a good agreement between adjusted and theoretical parameters given in **Table 1**.

The agreement between fitted and theoretical values affirms the coupling between spins indicated above.

From the formula $\lambda = \frac{3k_BT_C}{NJ(J+1)g^2\mu_B{}^2}$ and $M_0 = NJg\mu_B$ and there adjusted values, we can estimate the value of the spin number N:

$$\begin{cases} N = \dfrac{3k_BT_C}{\lambda J(J+1)g^2\mu_B{}^2} = \dfrac{3 \times 1,30807 \times 131.10^{-23}}{0,4024 \times 5,603 \times (5,603+1) \times 2,498^2 \times (9,274.10^{-24})^2} \approx 6.10^{23} & \text{(a)} \\ N = \dfrac{M_0}{Jg\mu_B} = \dfrac{83,592}{5.603 \times 2,6432 \times 9,274.10^{-24}} \approx 6.10^{23} & \text{(b)} \end{cases}$$

The two equalities, (a) and (b), practically give the same spin number N witch verifying the validity of the mean field theory. In addition, the value of N is near to Avogadro number N_A. This implies that we can assume that molecule may be present in a same value of spin so an important order domain and the nonmagnetic molecules (impurities are very limited).

After injecting adjusted parameters λ, J, g, and M_0 in Eq. (6), we can get simulated *M* versus *H* curves (red lines), which are plotted with the experimental

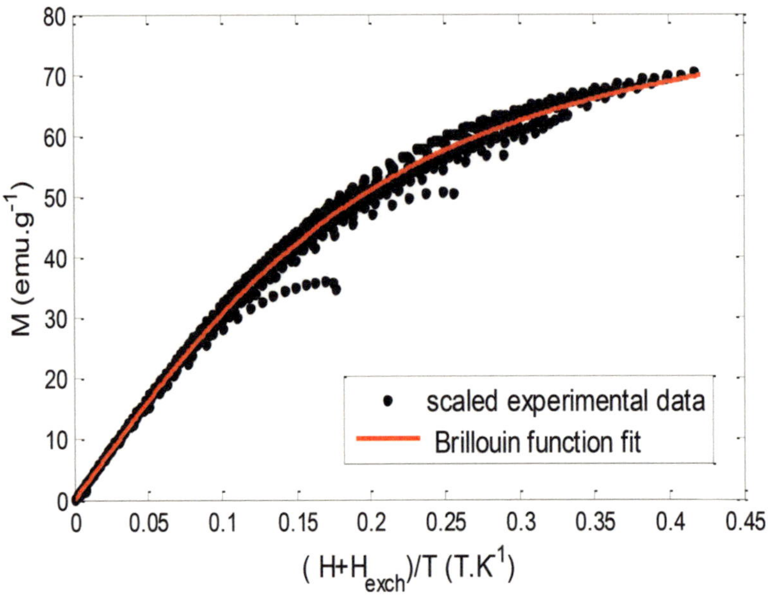

Figure 7.
Scaled data in magnetization versus $\frac{H+H_{exch}}{T}$ and Brillouin function fit for $Nd_{0.67}Ba_{0.33}Mn_{0.98}Fe_{0.02}O_3$ sample.

Parameters	J	g	M_0 (emu.g^{-1})
Theoretical values	5.614	2.643	—
Adjusted values	5.603	2.686	83.59

Table 1.
Theoretical and adjusted parameters of $Nd_{0.67}Ba_{0.33}Mn_{0.98}Fe_{0.02}O_3$ sample.

ones (black symbols) in **Figure 8**. This figure shows a good agreement between theoretical an experimental results. This illustrates the validity of the mean field to model the magnetization. On the other hand, **Figure 9** shows that simulated *M* versus *T* curves (red line) under various *H* are correlated with experimental ones

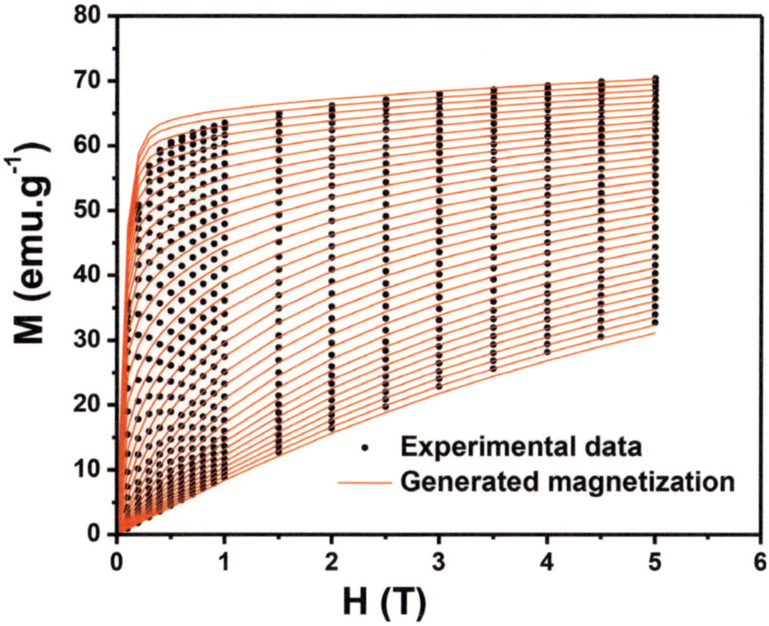

Figure 8.
Experimental M versus H (black symbols) of $Nd_{0.67}Ba_{0.33}Mn_{0.98}Fe_{0.02}O_3$ sample and the interpolation using the mean field method (red lines).

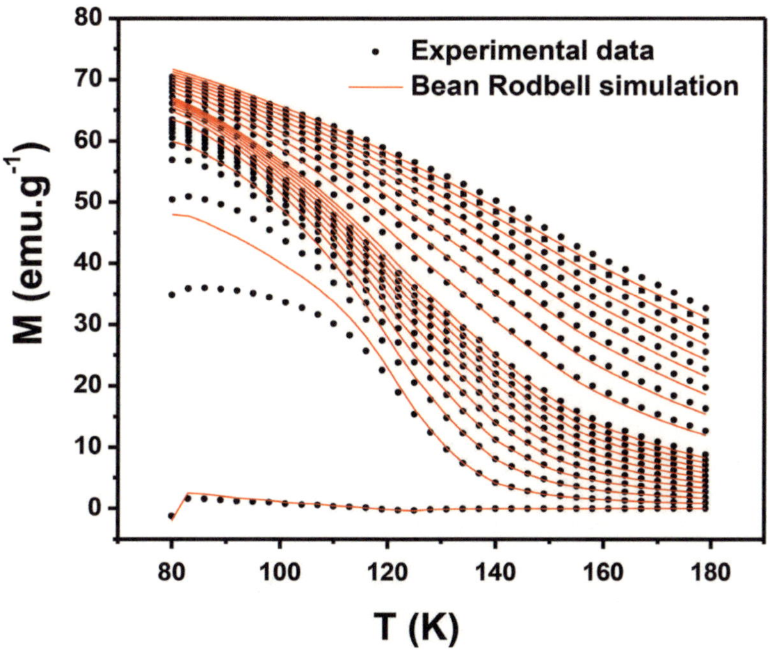

Figure 9.
Experimental magnetization versus T (black symbols) of $Nd_{0.67}Ba_{0.33}Mn_{0.98}Fe_{0.02}O_3$ sample and the interpolation using the Bean-Rodbell model (red lines).

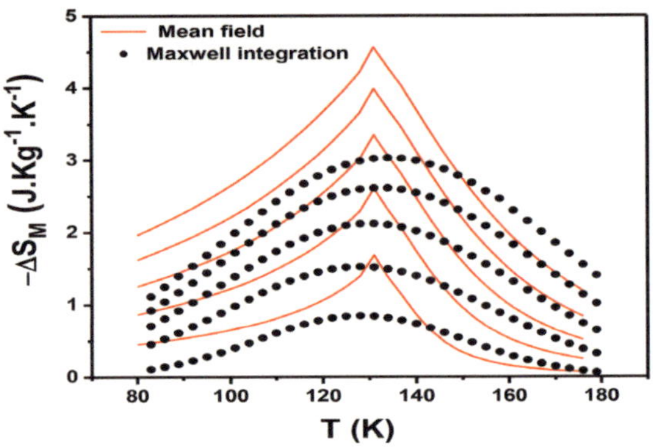

Figure 10.
Experimental and theoretical magnetic entropy change $-\Delta S_M$ versus T of $Nd_{0.67}Ba_{0.33}Mn_{0.98}Fe_{0.02}O_3$ sample as a function of temperature upon different magnetic field intervals (ΔH).

Method	H (T)	1	2	3	4	5
Maxwell relation	δT_{FWHM} (k)	47.54	54.34	56.98	59.15	63.01
	$-\Delta S_M^{max}$ (J.Kg^{-1}.K^{-1})	0.85	1.53	2.12	2.62	3.04
	RCP(J.Kg^{-1})	40.41	83.14	120.80	154.97	191.55
Mean field theory	δT_{FWHM} (k)	27.28	33.31	39.83	41.51	44.18
	$-\Delta S_M^{max}$ (J.Kg^{-1}.K^{-1})	1.65	2.59	3.32	3.98	4.54
	RCP(J.Kg^{-1})	45.01	86.27	132.24	165.2	200.58

Table 2.
Comparative between δT_{FWHM}, $-\Delta S_M^{max}$, and RCP calculated graphically using Maxwell relation and mean field theory.

(black symbols) when $\eta = 0.32$ and $T_0 = 131$ K. Thus, the second-order phase transition of this compound is reconfirmed with the η parameter value ($\eta < 1$).

Figure 10 shows simulated $-\Delta S_M$ versus T curves (red lines) using Eq. (9b) and the experimental ones (black symbols) using the Maxwell relation from in Eq. (9a). As seen in this figure and taking account into the initial considering of H and M as an internal and external variable in Eq. (8a) and vice versa in Eq. (8b), $-\Delta S_M$ estimated in these two considerations is little different. This aspect has been reported in the work of Amaral et al. [18]. From **Figure 10**, we can estimate the full width at half maximum δT_{FWHM}, the maximum magnetic entropy change $-\Delta S_M^{max}$, and the relative cooling power (RCP) which is the product of $-\Delta S_M^{max}$ and δT_{FWHM}. These magnetocaloric properties are listed in **Table 2**.

As shown in **Table 2**, a rising of $-\Delta S_M^{max}$ obtained by using the mean field model could be noted. For example, it exceeds the one determined by using the classical Maxwell relation by 1.5 J.Kg^{-1}.K^{-1} under 5 T applied field. Although δT_{FWHM} determined by this method seems less, RCP values are more higher than those obtained from the Maxwell relation. As a result, the mean field model could amplify RCP. This novel method has so better performance than the classical Maxwell relation.

Considering the number of magnetic near neighbors ions, z, in our material and its critical temperature, the relation $J = \frac{3k_B T_c}{2zJ(J+1)}$ (Eq. (26)) allows us to find the Heisenberg exchange constant J. In the perovskite structure of $Nd_{0.67}Ba_{0.33}Mn_{0.98}Fe_{0.02}O_3$ compound, the Mn ion placed at the center of the pseudo-cubic cell has four near neighbors Nd distant from $\frac{a\sqrt{3}}{2}$ and six near neighbors Mn distant from a and similarly for Nd. The interaction is established between Mn-Mn and Mn-Nd or Nd-Nd and Nd-Mn.

By averaging these interactions, the relationship (19) should be written as

$$J = \frac{3k_BT_c}{2} \times \frac{1}{\left[\frac{z_{Mn}S(S+1)+z_{Mn-Nd}J_{Mn-Nd}(J_{Mn-Nd}+1)+z_{Nd}J(J+1)}{3}\right]}$$

where $z_{Mn-Mn} = 6$, $z_{Mn-Nd} = 4$, and $z_{Nd-Nd} = 6$

$$S(Mn, Fe) = S(Mn^{3+}) + S(Mn^{4+}) - S(Fe^{3+})$$
$$= 0.65 \times 2 + 0.33 \times 1,5 - 0,02 \times 2.5 = 1.3 + 0.495 - 0.05 = 1.745;$$

$J_{Nd} = 3.855$; $J_{Mn-Nd} = 5.66$. So, $J = \frac{3\times 1.3807\times 10^{-23}\times 131}{2} \times \left[\frac{1}{\frac{6\times 1.745\times 2.745+4\times 5.6\times 6.6+6\times 3.855\times 4.855}{3}}\right]$ $= 2.8175 \times 10^{-23}$ joules for magnetic ion in our sample. This value explains the strength interaction between spins. Moreover, it is a crucial parameter used in the simulation with the Monte Carlo method.

3. Conclusion

In this work, we have analyzed the mean field scaling method for the $Nd_{0.67}Ba_{0.33}Mn_{0.98}Fe_{0.02}O_3$ sample. The perspicacity saved from the usefulness of this method for a magnetic system could be of large interest. In a simple reason, we can consider that if this scaling method does not follow the mean field behavior, other methods need to be convinced in the interpretation of the system's magnetic behavior. The mean field scaling method allows us to estimate the exchange parameter λ, the total angular momentum (J), the gyromagnetic factor g, the number of spins N of our sample, the saturation magnetization M_0, and the Heisenberg exchange constant **J**. Some of these factors are useful in estimating some magnetic properties. The mean field and the Bean-Rodbell models allow to follow the evolution of generated magnetization curves as function as the applied field and the temperature. A good agreement between theoretical and experimental magnetizations has been noted. The dependence of the entropy change on temperature under various applied fields has been experimentally and theoretically derived. An acceptable agreement between theoretical and experimental results is observed. However, the performance of RCP has been granted by the mean field model. Also, intervention of the Bean-Rodbell model confirms the second-order magnetic transition of our sample. Because this type of transition is needed for evaluating the MCE, a significant theoretical description of magnetic and magnetocaloric properties of the $Nd_{0.67}Ba_{0.33}Mn_{0.98}Fe_{0.02}O_3$ sample should be taking into account and should be accordable with other models.

Author details

Mohamed Hsini* and Sadok Zemni
Faculty of Science Monastir, Sidi Bouzid, Tunisia

*Address all correspondence to: mohamed.hsini.14@gmail.com

References

[1] Phan MH, Yu SC. Review of the magnetocaloric effect in manganite materials. Journal of Magnetism and Magnetic Materials. 2007;**308**:325-340. DOI: org/10.1016/j.jmmm.2006.07.025

[2] Kitanovski A, Egolf PW. Application of magnetic refrigeration and its assessment. Journal of Magnetism and Magnetic Materials. 2009;**321**:777-781. DOI: 10.1016/j.jmmm.2008.11.078

[3] Ke YJ, Zhang XQ, Ma Y, Cheng ZH. Anisotropic magnetic entropy change in $RFeO_3$ single crystals (R = Tb, Tm, or Y). Chinese Physics B. 2015;**24**:037501. DOI: 10.1038/srep19775

[4] Wada H, Tanabe Y. Giant magnetocaloric effect of $MnAs_{1-x}Sb_x$. Applied Physics Letters. 2001;**79**:3302. DOI: 10.1063/1.1419048

[5] Balli M, Fruchart D, Gignoux D. A study of magnetism and magnetocaloric effect in $Ho_{1-x}Tb_xCo_2$ compounds. Journal of Magnetism and Magnetic Materials. 2007;**314**:16-20. DOI: 10.1016/j.jmmm.2007.02.007

[6] Tishin AM, Spichkin YI. The Magnetocaloric Effect and Its Applications. London: IOP Publishing; 2003

[7] Dong Q, Zhang H, Sun J, Shen B, Franco V. Magnetocaloric response of $Fe_{75}Nb_{10}B_{15}$ powders partially amorphized by ball milling. Journal of Applied Physics. 2008;**103**:116101. DOI: 10.1063/1.3155982

[8] De Oliveira N. Magnetocaloric effect in the pseudobinaries (Ho1−cRc)Co_2 (R = Er and Dy). Eur. Phys. J. B. 2008;**65**:207-212. DOI: 10.1140/epjb/-2008-00346-y

[9] Wang ZW, Yu P, Cui YT, Xia L. Near room temperature magneto-caloric effect of a $Gd_{48}Co_{52}$ amorphous alloy. Journal of Alloys and Compounds. 2016; **658**:598-602. DOI: 10.1016/j.jallcom.2015.10.293

[10] Hamad MA. Magnetocaloric effect in $La_{0.7}Sr_{0.3}MnO_3/Ta_2O_5$ composites. J. A. D. 2013;**2**:213-217. DOI: 10.1007/s40145 -013-0062-0

[11] Phejar M, Paul-Boncour V, Bessais L. Investigation on structural and magnetocaloric properties of $LaFe_{13-x}Si_x(H,C)_y$ compounds. Journal of Solid State Chemistry. 2016;**233**:95-102. DOI: org/10.1016/j.jssc.2015.10.016

[12] Tlili R, Omri A, Bejar M, Dhahri E, Hlil EK. Theoretical investigation of the magnetocaloric effect of $La_{0.7}(Ba, Sr)_{0.3}$ MnO_3 compound at room temperature with a second-order magnetic phase transition. Ceramics International. 2015; **41**:10654-10658. DOI: 10.1016/j.ceramint.2015.04.165

[13] Tlili R, Hammouda R, Bejar M, Dhahri E. Effect of Ga substitution on magnetocaloric effect in $La_{0.7}(Ba, Sr)_{0.3}$ $Mn_{1-x}Ga_xO_3$ $(0.0 \leq x \leq 0.20)$ polycrystalline at room temperature. Journal of Magnetism and Magnetic Materials. 2015;**399**:143-148. DOI: 10.1016/j.jmmm.2015.09.073

[14] Raju K, Pavan Kumar N, Venugopal Reddy P, Yoon DH. Influence of Eu doping on magnetocaloric behavior of $La_{0.67}Sr_{0.33}MnO_3$. Physics Letters A. 2015;**379**:1178-1182. DOI: 10.1016/j.physleta.2015.02.016

[15] Coey J. Magnetism and Magnetic Materials. Cambridge: Cambridge University Press; 2009

[16] Kittel C. Introduction to Solid State Physics. 7th ed. New York: Wiley; 1996

[17] Gonzalo JA. Effective Field Approach to Phase Transitions and Some Applications to Ferroelectrics. Singapore: World Scientific; 2006

[18] Amaral JS, Das S, Amaral VS. The meanfield theory in the study of ferromagnets and the magnetocaloric effect, thermodynamics —systems in equilibrium and nonequilibrium. In: Piraj'a JCM, editor. Rijeka, Croatia: InTech; 2011. Available from: https://www.intechopen.com/books/thermodynamics-systems-in-equilibrium-and-non equilibrium/th0.33emean-field-theory-in-the-study-of-ferromagnets-and-the-magnetocaloric-effect

[19] Hcini S, Boudard M, Zemni S, Oumezzine M. Effect of Fe-doping on structural, magnetic and magnetocaloric properties of $Nd_{0.67}BaMn_{1-x}Fe_xO_3$ manganites. Ceramics International. 2014;**40**:16041-16050. DOI: 10.1016/j.ceramint.2014.07.140

[20] Dong QY, Zhang HW, Sun JR, Shen BG, Franco V. A phenomenological fitting curve for the magnetocaloric effect of materials with a second-order phase transition. Journal of Applied Physics. 2008;**103**:116101-1161015. DOI: 10.1063/1.2913166

[21] Bean CP, Rodbell DS. Magnetic disorder as a first-order phase transformation. Physics Review. 1962; **126**:104-115. DOI: 10.1103/PhysRev.126.104

[22] Baazaoui M, Boudard M, Zemni S. Magnetocaloric properties in $Ln_{0.67}Ba_{0.33}Mn_{1-x}Fe_xO_3$ (Ln=La or Pr) manganites. Materials Letters. 2011;**65**: 2093-2095. DOI: 10.1016/j.matlet.2011.04.051

[23] Mostafa AG, Abdel-Khalek EK, Daoush WM, Moustafa SF. Study of some co-precipitated manganite perovskite samples-doped iron. Journal of Magnetism and Magnetic Materials. 2008;**320**:3356-3360. DOI: 10.1016/j.jmmm.2008.07.025

[24] Ogale SB, Shreekala R, Bathe R, Date SK, Patil SI, Hannoyer B, et al. Transport properties, magnetic ordering, and hyperfine interactions in Fe-doped $La_{0.75}Ca_{0.25}MnO_3$: Localization-delocalization transition. Physical Review B. 1998;**57**:7841. DOI: 10.1103/PhysRevB.57.7841

[25] Callen HB. Thermodynamics and an Introduction to Thermostatistics. 2nd ed. New York: Wiley; 1985

[26] Von Ranke PJ, de Oliveira NA, Gama S. Understanding the influence of the first-order magnetic phase transition on the magnetocaloric effect: Application to $Gd_5(Si_xGe_{1-x})_4$. Journal of Magnetism and Magnetic Materials. 2004;**2**(77):78-83. DOI: 10.1016/j.jmmm.2003.10.013

[27] Turcaud JA, Pereira AM, Sandeman KG, Amaral JS, Morrison K, Berenov A, et al. Spontaneous magnetization above T C in polycrystalline $La_{0.7}Ca_{0.3}MnO_3$ and $La_{0.7}Ba_{0.3}MnO_3$. Physical Review B. 2014;**90**:024410-024417. DOI: 10.1103/PhysRevB.90.024410

[28] Balli M, Fruchart D, Gignoux D, Tobola J, Hlil EK, Wolfers P, et al. Magnetocaloric effect in ternary metal phosphides $(Fe_{1-x}Nix)_2P$. Journal of Magnetism and Magnetic Materials. 2007;**316**:358-560. DOI: 10.1016/j.jmmm.2007.03.018

[29] Zach R, Guillot M, Tobola J. Semiquantitative analysis of magnetic phase transitions in the $MnFeP_{1-x}As_x$ series of compounds. Journal of Applied Physics. 1998;**83**:7237-7239. DOI: 10.1063/1.367856

[30] Jia L, Sun J, Zhang H, Hu F, Dong C, Shen B. Magnetovolume effect in intermetallics $LaFe_{13-x}Si_x$. Journal of Physics. Condensed Matter. 2006;**18**: 9999-10009. DOI: 10.1088/0953-8984/18/44/002

[31] Cyrot M. Magnetisme I—Fondements. Grenoble: Presses Universitaires de Grenoble; 1999

Chapter 3

The Magnetometry—A Primary Tool in the Prospection of Underground Water

Héctor López Loera

Abstract

One of the most important problems in arid and semi-arid zones in the Mexican Mesa Central is the one related to the exploration and exploitation of groundwater. It is found at depths over 200 m, and movement is primarily through fractures. This work presents a geophysical methodology, which shows the potential of combining natural and induced methods to locate confined aquifers in fault zones. The study begins by interpreting the aeromagnetic survey, mainly by searching alignments associated with low magnetic anomalies, which are correlated with faults zones, and/or fractures and/or geologic contacts where ferromagnetic minerals have undergone oxidation due to their association with recharged zones. These aero-magnetic alignments are confirmed on land by a ground magnetic survey. Based on these interpretations, electrical methods include sections and vertical electrical sounding are used to verify if the zones are correlated to the underground moisture. If both permeability and moisture are met together, then they considered as zones with a high probability of locating ground water in the Mexican Mesa Central.

Keywords: fractured-media aquifers, geophysical methods, aeromagnetics, semi-arid áreas, Mexican Mesa Central, San Luis Potosi

1. Introduction

Localization of ground water is a national and global issue at the priority basis. The fragility of water for various uses is a serious problem, which can be resolved by geological and Vertical Electric Soundings (VES) studies [1–3]. The problem arises when the geological information is not enough and accurate in the location of wells, resulting into dry wells. This problem is mostly frequent in volcanic areas, where the areas covered by alluvial material do not allow to observe the possible structures that contain underground water. In this study, we present a methodology for the location of this resource in arid volcanic zones, especially in the Central Mesa of Mexico. The methodology is based on a basic knowledge of Geology, the study of the magnetic field (air and ground) and the application of the electrical resistivity method, in two modalities, that is, sections and SEV [2].

The methodology was applied to solve serious water problem in the rural population of La Dulcita town, Municipality of Villa de Ramos, which is located at the Northwest of the capital of San Luis Potosí and state of Zacatecas (**Figure 1**). The population of La Dulcita in 2005 was reported with 750 inhabitants [4] and their water was supplied by a single well located at 5 km South of La Dulcita town, with

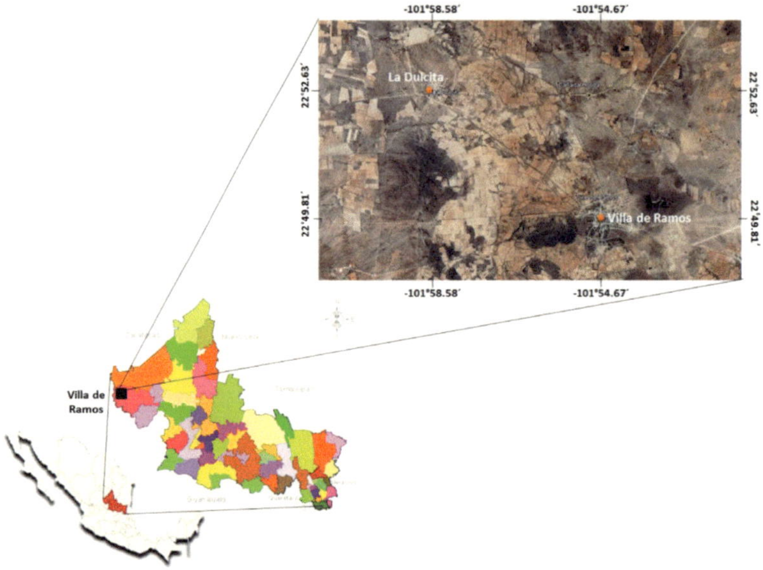

Figure 1.
Satellite image of study area, that is, La Dulcita, villa de Ramos, state of San Luis Potosí, Mexico.

its capacity measured less than 1 L/s, which was not sufficient for the entire population. In addition, the State Water Commission (CEA, for its acronym in Spanish), State of San Luis Potosí, had drilled three wells and all of them were dry.

The rocks that form the aquifers are characterized by their physical properties such as porosity, permeability and water content [5, 6]. The present methodology allows locating the zones and the degree of fracture and measure if these can be associated to moisture from the surface.

2. Study area

The Geology of the study area is represented mainly by the alluvial deposits approximately to the south of the La Dulcita, an outcrop of basaltic rocks exist in this area, whose height is approximately 15 m from the ground level (**Figure 2**). In the East, there are outcrops of the Caracol Formation, of the Upper Cretaceous [7] forming hills that protrude from the plains (**Figure 3**). It consists of shales of a greenish color, sometimes very dark gray. In the area of the Villa de Ramos, there is a large granite extension, which has almost a North-South course and constitutes a tectonic pillar that presents mineralization in some areas. In addition, also towards the North of Villa de Ramos, there are outcrops of marine sedimentary rocks [7].

La Dulcita area is located in a tectonic pit where the base must be represented by marine sedimentary rocks and probably basaltic lava flows.

3. Methodology

3.1 Geological survey

First, a compilation of the existing geological information that already exists of the State of San Luis Potosí that was published by the Institute of Geology, Autonomous University of San Luis Potosi and the Mexican Geological Service is performed. Once the existing information has been compiled, a geological survey

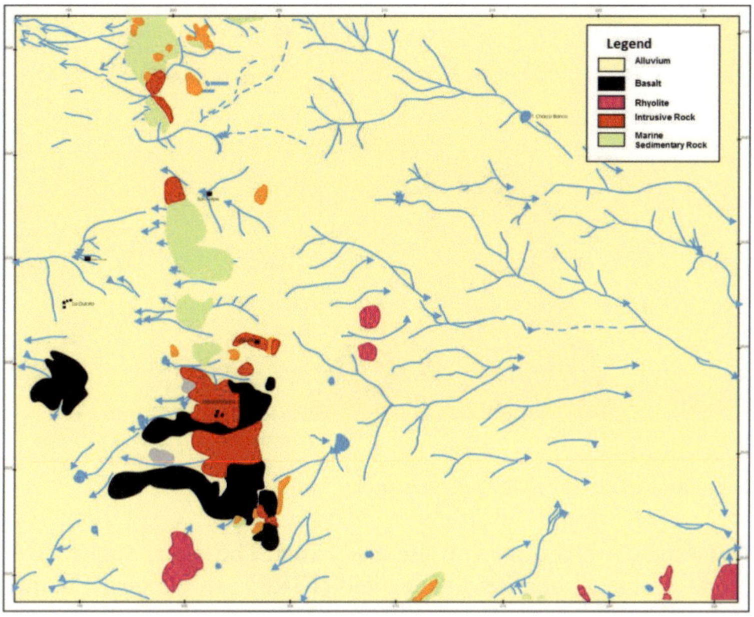

Figure 2.
Geological map of the Villa de Ramos area modified after Labarthe and Aguillón [7].

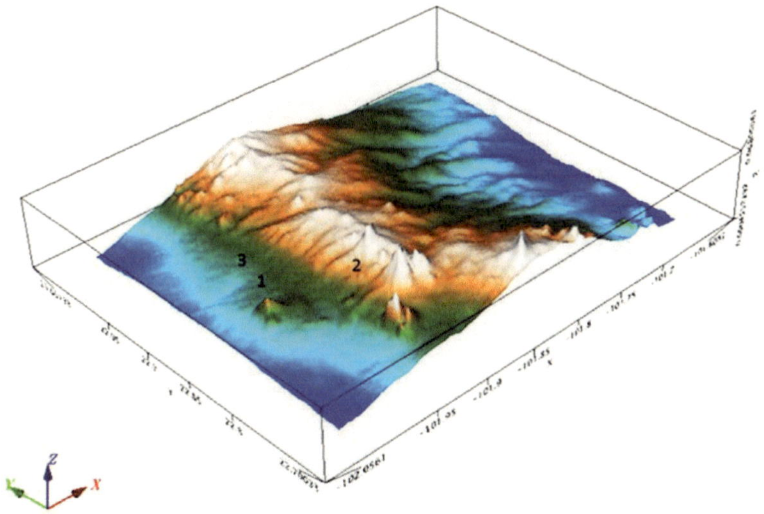

Figure 3.
Elevation digital model where La Dulcita (1) Villa de Ramos (2) and the H_2O well (3) are located, San Luis Potosí, Mexico.

of the study area is carried out to locate the geological units that can exist in the area under the study and a digital elevation model formed (**Figure 3**). In addition, **Figure 4** indicates a geological map of the study area and an idealized diagrammatic model where the main structures and existing geological units are indicated.

3.2 Geophysical study

The geophysical study is comprised of several stages; first, the aeromagnetic information of study area is analyzed. This is done by applying a series of mathematical algorithms (filters) to the aeromagnetic data, which allow highlighting certain

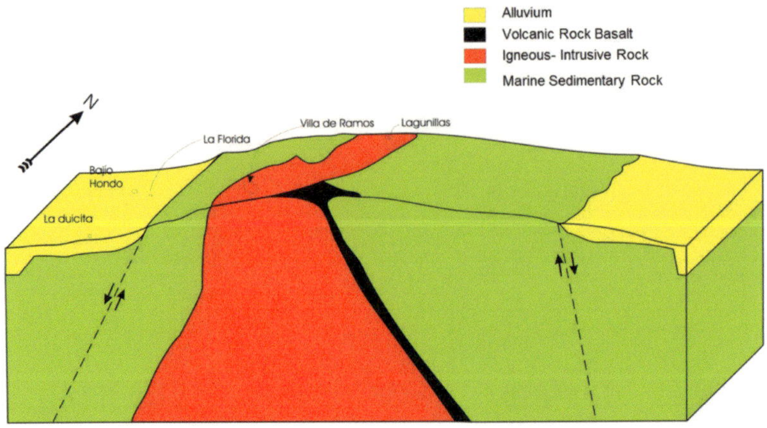

Figure 4.
Geological block of the La Dulcita-villa de Ramos area, San Luis Potosí, Mexico.

characteristics and dismiss others on the study area. The filters applied are known as International Geomagnetic Reference Field (IGRF) [8], which is calculated every 5 years and the immediate inferior should be applied to the date of the aerial survey (e.g., the aeromagnetic flight of our study area was carried out in 1995, the IGRF must be subtracted is that of 1990) [9]. To obtain the intensity values of the total magnetic field (TMF), which are obtained when flying, the contribution of the main dipole is subtracted, which exists in the terrestrial nucleus, thus obtaining the values of the residual magnetic field [RMF, Eq.(1)].

$$TMF - IGRF = RMF \tag{1}$$

Since the magnetic field is a vector (defined by magnitude and direction), the magnetic anomalies in these latitudes are displaced from the sources that produce them. Therefore, this is the reason why other mathematical algorithms must be applied for a filter, which simulates our study area, where the magnetic inclination is 90° and the declination is 0°. This algorithm named Baranov and Naudy [10] is better known as reduction to the magnetic pole field (RMPF) and assures us theoretically that magnetic anomalies will be located in the sources that produce them. The data matrix, thus generated, is the basis for the application of other filters or mathematical algorithms.

In arid volcanic areas, one of the opportunities to locate groundwater is in the confined aquifers on faults. A filter is a mathematical tool to guide us, if we want to know the fracture, faults or the contacts zones in the geological units a filter that provides us with guidance is the Henderson and Zietz [11]. This filter is known as vertical derivatives of first or second degree, because it is going to indicate the areas of high gradients which are normally associated with the geological structures mentioned above. Another filter that has been applied to aeromagnetic information is the Henderson [12] that allows us to change the plane of observation, when we rise, the high frequencies tend to attenuate and highlight the low frequencies, which are associated with the geological structure of the subsoil. This filter is known as the magnetic field upward continuation.

It is possible to interpret the location of the superficial and deep permeability zones with the analysis of the magnetic contour maps through each of these filters [13, 14].

The next step of the methodology, after the aeromagnetic information has been analyzed, is to perform a land magnetometry survey in the areas that have presented some possibility of being associated with fracturing and/or faulting and/or geological contacts [15, 16]. This stage is called anomaly verification.

The orientation of the land magnetic profiles should be as far as possible perpendicular to the structure that is inferred. The processing and analysis of the ground magnetic information is similar to the one made to the aeromagnetic data, a RMF is obtained from the TMF, later RMPF is generated and to this field, the filters of derivatives and upward continuation are applied.

The magnetic information analysis is up to a one-point simple. This method is based on the fact that the whole Earth behaves like a large magnet that would be in the center of it [17, 18]. For a specific area, it is considered that the magnetic field strength (**H**) is the same, and that the value of the magnetic intensity will be a function of the magnetic susceptibility of the rocks [*k*; Eq. (2)], which is defined as the capacity of these to acquire magnetization, such that:

$$I = k\mathbf{H} \tag{2}$$

This magnetization (I) constitutes the induced magnetization. Additionally, the effects of the remnant magnetization are present. In volcanic and intrusive rocks, this magnetization can be of greater intensity than the induced effects [18, 19]. If we consider that volcanic rocks contain ferromagnesian and is without fractures, it will generate a magnetic response characterized by having an anomaly represented by a magnetic high and a low. That is, the magnetic response has a positive and a negative side. If we make the simile that the rock is a magnet, and fractured it in two parts, we would generate two anomalies, which would have two magnetic highs and lows in a sequence, high-low-high-low magnetic, and so on. If we have a slightly fractured and/or faulty area, it will give us a magnetic response with highs and lows sequenced, with medium frequencies. In contrast, if we have a zone with highly fractured and/or high faulting, it will give us a magnetic response characterized by high frequencies and by a series of sequenced of magnetic highs and lows [16]. These areas are interpreted as zones where two of three of the properties that identify the aquifers are present, as they have porosity and permeability.

Once the secondary permeability zone in the volcanic rocks has been identified, the next step in the methodology is to prove that the zones associated with humidity, which is achieved with the application of electrical methods; in our case, we use direct current.

The electric DC methods are based on Ohm's law [2], which establishes that the resistance is directly proportional to the voltage and inversely proportional to the intensity of the electric current:

$$R = \frac{V}{I} \tag{3}$$

where R (Ohm, Ω) is the electrical resistance, V (Volts) the potential and I (Amp) in the electric current.

The previous relationship (Eq. 3) is valid for any electrical circuit, in studies of the underground, the relationship changes since the resistance is a function of the nature and the geometry of the conductor (Eq. 4), in this case the Earth:

$$R = \frac{\rho L}{S}, \ \rho = \frac{RS}{L} \tag{4}$$

The equation in which ρ represents the nature of the conductor and is called resistivity, L is the length in m and S the conductor section in m^2. If we replace and isolate the variable, then the following equation is given by:

$$\rho = \frac{VS}{IL} \tag{5}$$

In geoelectrical exploration, the resistivity of the underground is normally measured with an electrode arrangement of four electrodes, with the electrodes AB being the emission electrodes (current) and MN the potential electrodes.

In this case, the resistivity (Eq. 6) is given by:

$$\rho a = \frac{V}{I} \times 2\pi K \tag{6}$$

where K is the geometric factor (1/AM-1/BM-1/AN+1/BN) of the electrode array, the subscript "a" in the resistivity indicates that the calculated value is apparent.

The resistivity is an inverse property of the electrical conductivity and in exploration, its units is ohm per meter (Ω/m).

In geoelectric exploration, the variation of the resistivity is studied horizontally by means of profiles in which the electrode array is moved as a whole to the different stations. Conversely, equispaced or the vertical variation of the resistivity can be studied by means of vertical electric soundings (VES). At a certain point, for this, the current electrodes (AB) are increasingly opened and the measuring or potential electrodes (MN) are opened only when the measured values are very small (Schlumberger electrode array). In such manner in which data exist in one or two points with different MN opening for the same values of AB, there is an overlap or "cluth" during the measurement of the SEV [1].

A quick way to know the electrical behavior of the underground in a given area is to make profiles of electrical resistivity of two electrode openings, for example, at 200 and 400 m opening of the current electrodes with the Schlumberger electrode array (AB/2 at 100 and 200 m). In this way, we have knowledge of the variation of electrical resistivity in a horizontal direction.

If the resistivity behaves similar to both electrode array separations, it will imply that the entire electrically scanned area is the same. If the resistivity of the profile generated with a larger electrode aperture is higher than that generated at a smaller aperture, it will indicate that at depth the possibility of detecting humidity is zero. On the other hand, if the resistivity is lower at larger electrode aperture, it will have greater chances of detecting humidity. If the profiles of apparent resistivity show an irregular behavior, it will have greater possibilities of detecting a resistive resistance where the resistivity at the largest aperture changes form more resistive to less resistive and indicates that the area under study shows in the underground, where the current circulates more easily and will be an area where the variation of resistivity with depth must be studied, which is done with vertical electric sounding (SEV, [20]).

The VES' must be interpreted qualitatively and quantitatively. Firstly, the morphology of the VES curve must be defined [21] which in order to be associated with humidity, must necessarily have a correlation with the H-type curves ($\rho 1 > \rho 2 < \rho 3$), which indicates that there is a lower resistivity contrast between the central layer and those that enclose it. The VES curves can also be KH ($\rho 1 < \rho 2 > \rho 3 < \rho 4$), QH ($\rho 1 > \rho 2 > \rho 3 < \rho 4$) or some of the curves that show a portion of type H.

The quantitative interpretation is carried out using commercial software that allows an inversion of the resistivity data [22]. It is convenient to perform a VES in wells where its stratigraphic column is known, in a way that the VES can be calibrated.

Once the previous stages have been carried out, the zones that are chosen for drilling must have a magnetic response which correlates with a fractured zone (permeability) and electrical methods (resistivity) with an area that has a relation with a humid area, represented by a resistive contrast that contains a minimum between two resistivity maxima.

4. Results

The procedure described above has been applied to an area, which is located in the Mesa Central, Mexico, specifically to a rural population called La Dulcita, municipality of Villa de Ramos, San Luis Potosí.

4.1 Air Magnetometry

The area under study was flown by the Mexican Geological Service, using an Islander aircraft BN2-A21, equipped with a Geometrics G-822 magnetometer, of cesium vapor optical pump, with a sensitivity of 0.25 nT, and an acquisition system of P-101 Picodas data, Automax video camera, 35 mm. A Geometrics G-826A magnetometer was used, with a sensitivity of 1 nT as the base station. Also, Sperry altimeter radar was also used.

The course of the flight lines was N-S, with a distance between flight lines of 1000 m and a height above ground level of 300 m, the navigation was controlled with an Ashtech GG24 GPS system and the data was subtracted from the IGRF 1990 reference.

The total magnetic field intensity in the central portion was 44,858 nT, with an inclination of 50°43′ and declination of 8°13′ for July 1995.

The magnetic field behavior analysis began with the generation of the RMF map (**Figure 5**), which as mentioned in previous paragraphs, is obtained by subtracting the IGRF from the total magnetic field. Based on the RMF, the RMPF was calculated (**Figure 6**). In the W portion of the RMPF, there is a "trend" of magnetic highs (red) that represent the W limit of an area of the graben that exists with a general direction N-S and is characterized on the map with anomalies associated with magnetic lows (blue color). Towards the central portion, two "trends" of magnetic anomalies with direction NE–SW and NNW–SSE are shown that are possibly associated with the geologically multiple intrusive "El Socorro" [7]. The Dulcita area is located on the first step of the graben and alignments (**Figure 7**) with direction N-S and E-W towards its portion W is observed, which can be geologically associated with zones of faults and/or fracturing and/or contacts. The area investigated in

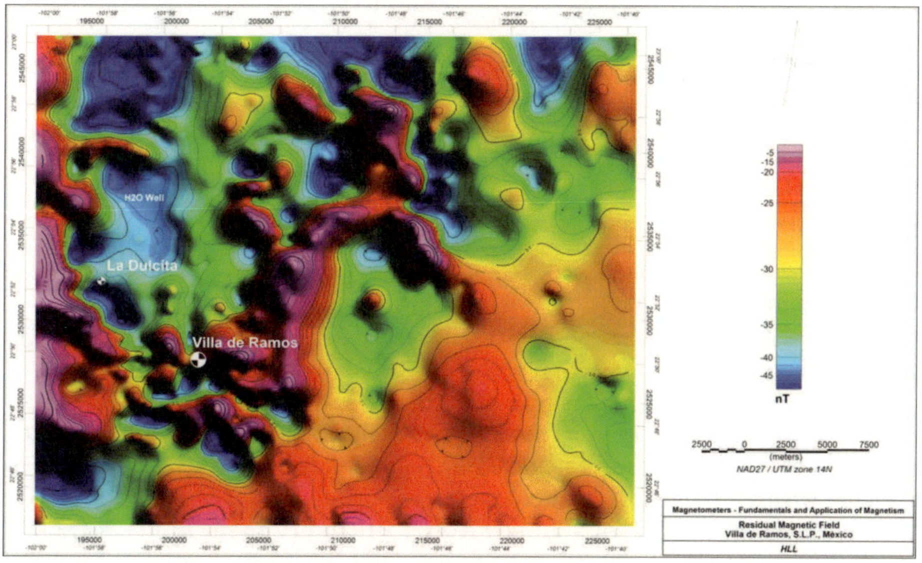

Figure 5.
Map showing the isovalues contour of the residual magnetic field of the Dulcita area, Villa de Ramos, San Luis Potosí, Mexico.

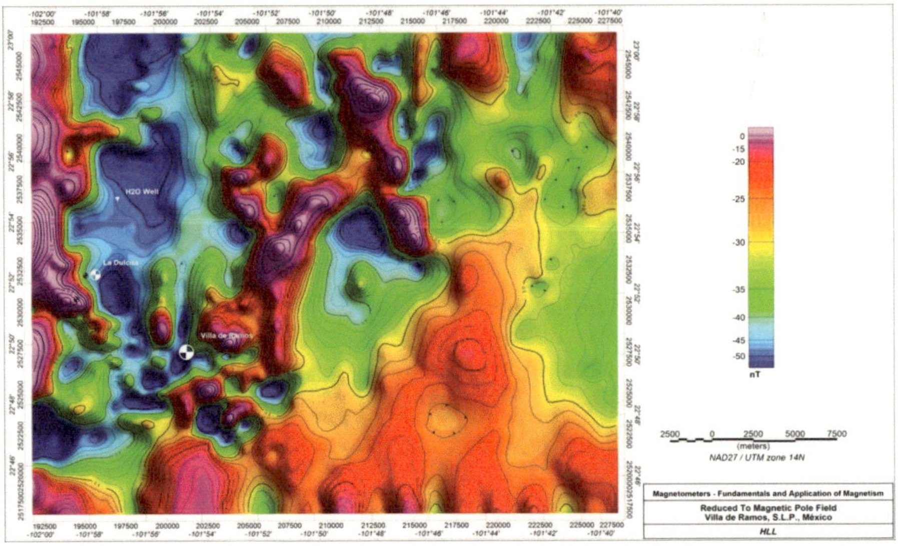

Figure 6.
Map showing the isovalues contour of the reduced to the pole magnetic field of La Dulcita, Villa de Ramos, San Luis Potosí, Mexico.

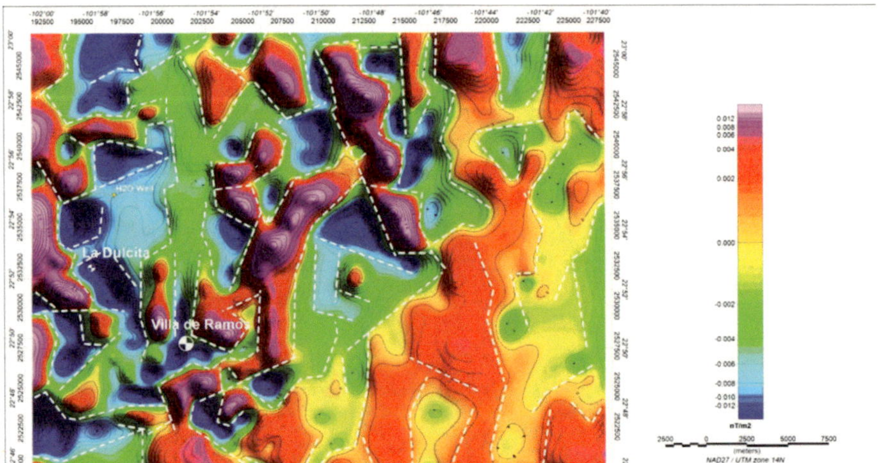

Figure 7.
Map where the magnetic alignments are observed based on the isovalues contour of the first vertical Derivativ upwards continuation 250 m from the reduced to the pole magnetic field.

general shows preferential aeromagnetic alignments in an N-S direction, also existing in the NE–SW direction, with few showing NW-SE direction.

The analyzed area in general shows the existence of up to 10 AMD's, each characterized by different amplitudes and wavelengths. The area where the water is extracted for the population of La Dulcita, is correlated with the AMD II that is associated with a tectonic pit area, characterized by low values of magnetism. The graben is limited by AMD I to W and by AMD's IIII and IV to the E. In AMD I, a highly productive well was located for the area (16 L/s) at a distance of 2.3 km SW of La Dulcita outside the ejido boundaries.

La Dulcita area is located in the aeromagnetic domain map (AMD), zone that show similar magnetic susceptibility (**Figure 8**) and is situated between the limits of AMD's I, II and IX, which allows us to interpret possibilities of the existence of permeability in the zones of the contacts.

4.2 Ground Magnetometry

From above interpretation of the aeromagnetic information, four ground magnetic sections were programmed with reading stations of the total magnetic field (TMF), during every 20 m, by using two magnetometers, one GEM-GSM-19 and another Geometrics G-856 A, to perform the measurements, in which they were corrected by daily and hourly drift and a residual was obtained by subtracting a zero-degree polynomial from the TMF.

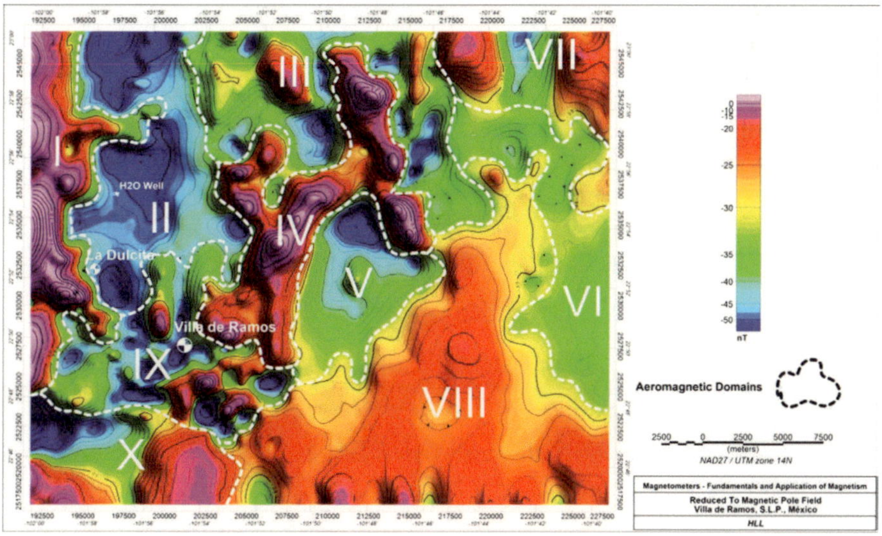

Figure 8.
Map of the aeromagnetic domains (AMDs) interpreted in the isovalues contour of the magnetic reduced pole field.

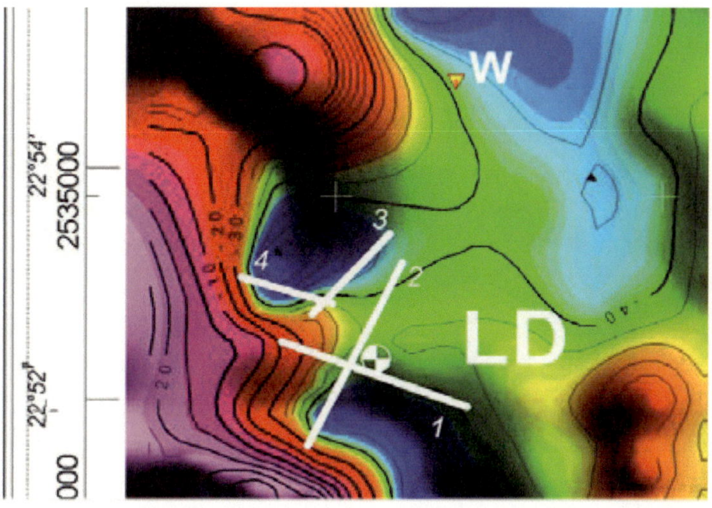

Figure 9.
Map showing the location of the ground magnetic sections. The water well that appears to the north of the map where the population of La Dulcita is supplied with a yield of less than 1 L/s.

Two of the sections had NW-SE orientation and two NE–SW (**Figure 9**) with the population of La Dulcita being in the central part of these profiles.

The magnetic section 1 (**Figure 10**) displays four terrestrial magnetic domains (TMD): first station 0 to 54 was characterized by a series of magnetic anomalies related to short wavelengths (20–40 m), high frequencies and amplitudes of 160 nT. It was geologically correlated with a highly fractured zone, while the horizontal gradients give values of up to 11 nT/m. Second, TMD 2 is located between stations 55 and 78, and it is defined by presenting a normal magnetic field, where no abnormal areas are observed. Third TMD 3 is located between stations 79 and 87 and shows an anomalous zone limited by two magnetic anomalies that have amplitudes of 33 and 65 nT and horizontal gradients of 2.6 and 6 nT/m, respectively. Geologically, it is correlated with an area of medium fracture possibilities. The last, TMD 4 is limited between stations 88 and 135, in general it shows a discretely disturbed magnetic field where it is not considered with the possibility of associating at depth with permeability.

The magnetic section 2 is located towards the E portion of La Dulcita (**Figure 9**), it presents five TMD's (**Figure 11**), the first one limited between stations 0 and 32 shows a normal behavior of the RMF, where magnetic anomalies are distinguished.

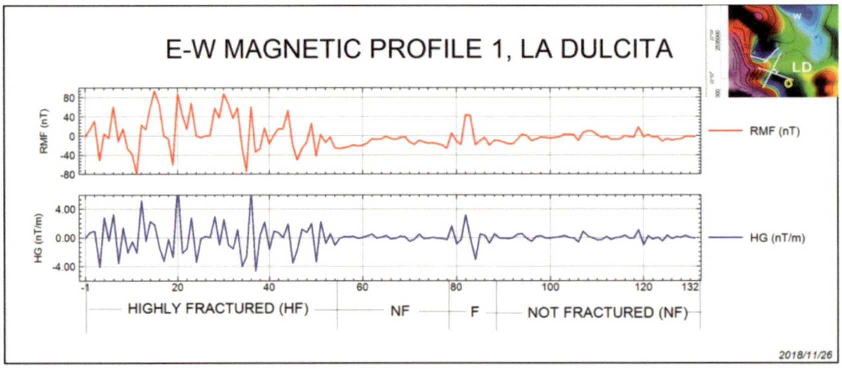

Figure 10.
Ground magnetic profile 1, with a NW-SE orientation. At the upper part, the residual magnetic field (RMF) is plotted (red); the horizontal gradient of the RMF is plotted at the lower part (blue), and at the bottom a qualitative interpretation of the percentage of probabilities of association with fracturing in the underground is shown. NF, not fractured; F, fractured.

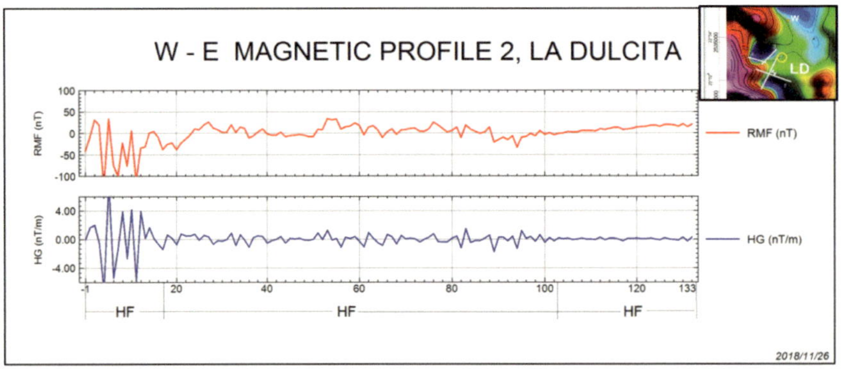

Figure 11.
Ground magnetic profile 2, with a NE–SW orientation. At the upper part the residual magnetic field (RMF) is plotted (red); the horizontal gradient of the RMF is plotted at the lower part (blue), and at the bottom a qualitative interpretation of the percentage of probabilities of association with fracturing in the underground is shown. HF, highly fractured; LF, light fractured; NF , not fractured.

The TMF 2 is located between stations 33 and 45 and does not show areas of high frequencies that can be correlated with fracturing effects at depth. The TMF 3 is located between stations 46 and 84, it is identified by presenting a magnetic response characterized by anomalies with short wavelengths (60–100 m), high frequencies and amplitudes of the order of 28–41 nT and horizontals gradients from 2.7 to 2.3 nT/m, respectively. It correlates an area with average possibilities that associating with the existence of secondary permeability. The TMD 4 is identified between stations 85 and 113 and has short wavelengths (20–80 m) high frequencies and amplitudes of 18 at 29 nT and horizontal gradients of 0.4–1.5 nT/m, respectively. They are geologically associated to an area with average possibilities of correlation with permeability in the underground. The TMD 5 is delimited between stations 114–133, characterized by showing short wavelengths (20–40 m), high frequencies and magnetization amplitudes of 54 nT up to 160 nT, with horizontal gradients of 6.5 nT/m up to 14.7 nT/m, is geologically correlated with an area of strong fracture and permeability.

The magnetic section 3 is located in the NW of La Dulcita (**Figure 9**) shows two TMD (**Figure 12**). The first domain is located between stations 1 and 16 is identified by presenting a series of magnetic anomalies. These are characterized by short wavelengths (20–100 m), high frequencies and amplitudes from 32 to 107 nT and horizontal gradients from 2.9 to 7.6 nT/m, which correlates with average possibilities of being associated in the underground with fracturing. The second TMD is located from station 17 to 75 and shows a normal magnetic field where the possibility to correlate with fracturing at depth zero.

The magnetic section 4 located outside La Dulcita, the NW portion (**Figure 9**), shows two TMD (**Figure 13**), neither of them of interest to be associated with fractured zones in the underground.

4.3 Electrical methods

Two electrical sections (or profiles) of apparent resistivity, induced polarization and self-potential were made with the Schlumberger type electrode array (**Figure 14**), using two electrode spacings AB/2 = 100 and 200 m and a Syscal R-2 resistivity instrumental (**Figure 15**). The sections were made in the same directions as the magnetic profiles 1 and 2, which were showed more possibilities of associating with fracturing in the underground.

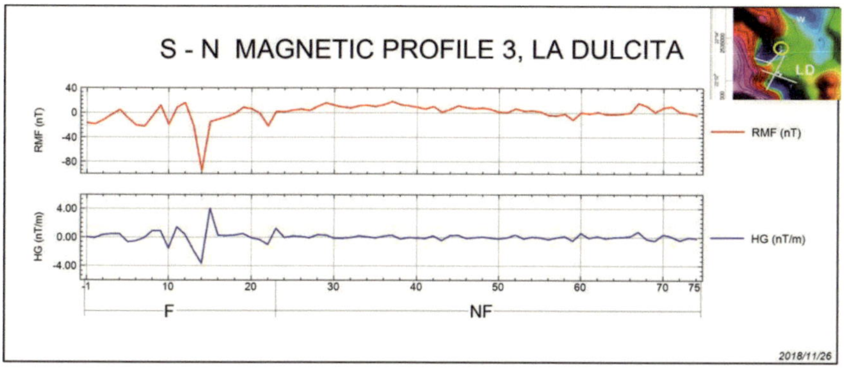

Figure 12.
Ground magnetic profile 3, with a NE–SW orientation. At the upper part the residual magnetic field (RMF) is plotted (red); the horizontal gradient of the RMF is plotted in the lower part (blue), and at the bottom qualitative interpretation of the percentage of probabilities of association with fracturing in the underground is shown. F, fractured; NF, not fractured.

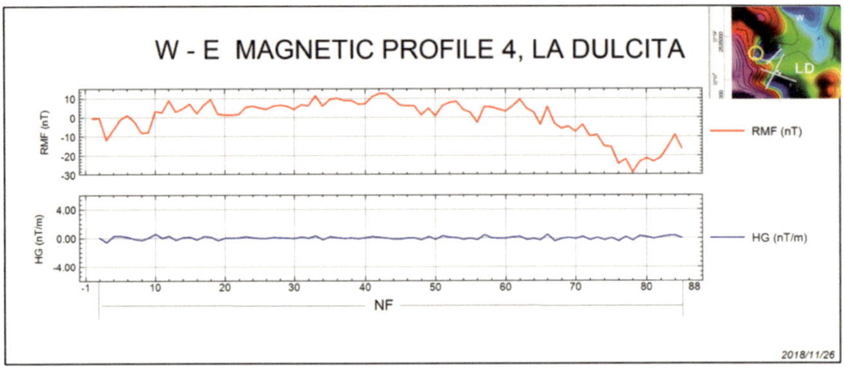

Figure 13.
Ground magnetic profile 4, with a NW-SE orientation. At the upper part the residual magnetic field (RMF) is plotted (red); the horizontal gradient of the RMF is plotted at the lower part and at the bottom a qualitative interpretation of the percentage of probabilities of association with fracturing in the underground is shown. NF, not fractured.

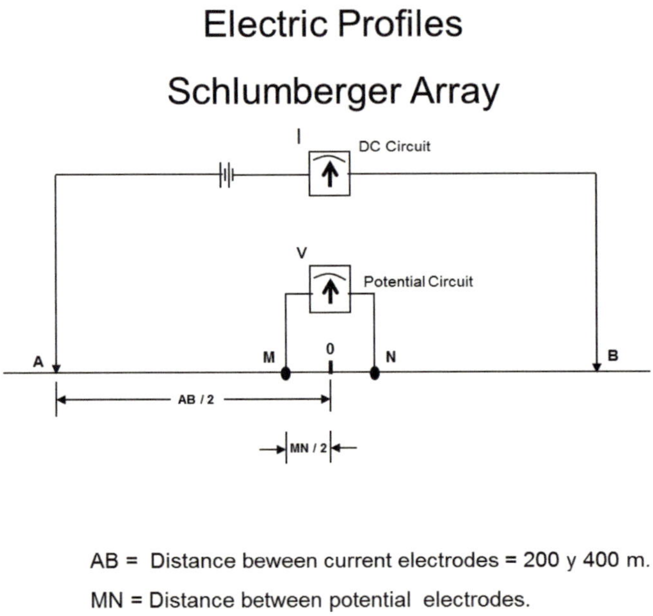

Figure 14.
The Schlumberger electrode array diagram, used for the realization of vertical electric sections and soundings (VES). The maximum openings of the VES were AB/2 of 1500 and 2000 m.

The W-E electrical profile shows in general an increase in resistivity with depth except two areas, from station 400 to 500 and 750 where the conductivity is higher. The induced polarization in these profiles generally shows a decrease in chargeability, except for two areas of station 450–500 and 750, where the load capacity tends to increase. The spontaneous potential is observed to decrease in general with larger separations of AB/2 (**Figure 16**).

The S-N electrical section, presents values of apparent resistivity lower at depth for the most part, except from station 450 to 550 where there is a small increase in resistivity to separations greater than AB/2. The chargeability values in the induced polarization are observed in contrast throughout the section, increase to greater separations of AB/2 in the areas of station 0–150, 350, 550–900 and in the station 1100. The spontaneous potential (SP) in this section behaves similarly to both

Figure 15.
Electrical instruments used for vertical electric soundings and sections.

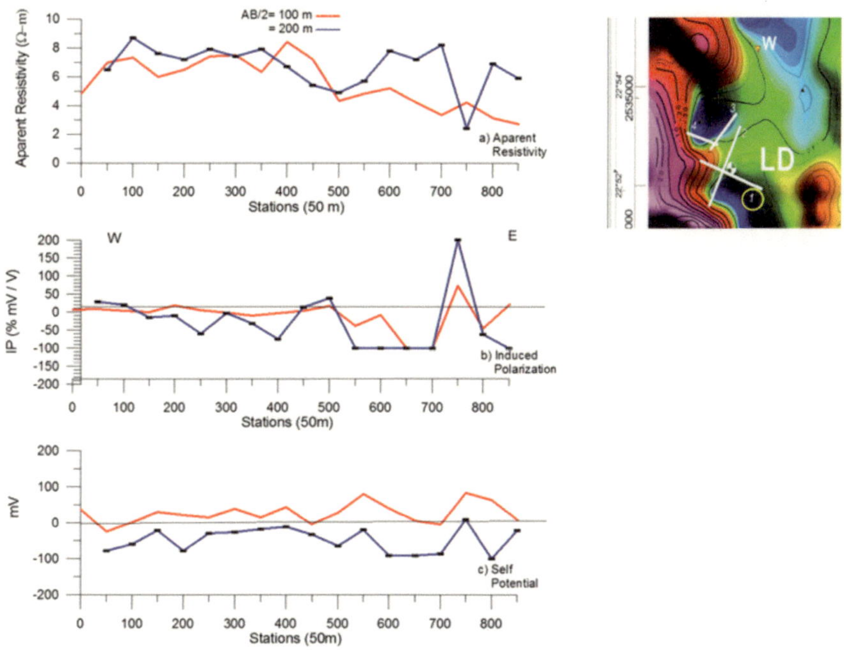

Figure 16.
Electrical profile 1, with a NW-SE orientation, where (a) the apparent resistivity is plotted; in (b) the induced polarization and in (c) the self-potential. These electrical profiles were located on the zones showing high frequencies (fracture, permeability) in profile 1 (Figure 10) of magnetometry.

electrode separations between stations 0 and 550, where at higher separations of AB/2, the values (mV) increase slightly from station 600 to 900, the values decrease for AB/2 = 200 m and from station 950 to 1300, the SP is changing (**Figure 17**).

4.3.1 Vertical electrical soundings

Five vertical electric soundings (VES') were made with maximum openings of the current electrodes (AB/2) of 1500 and 2000 m, four of them are located in identified zones (magnetometry) with possibilities of associating depth

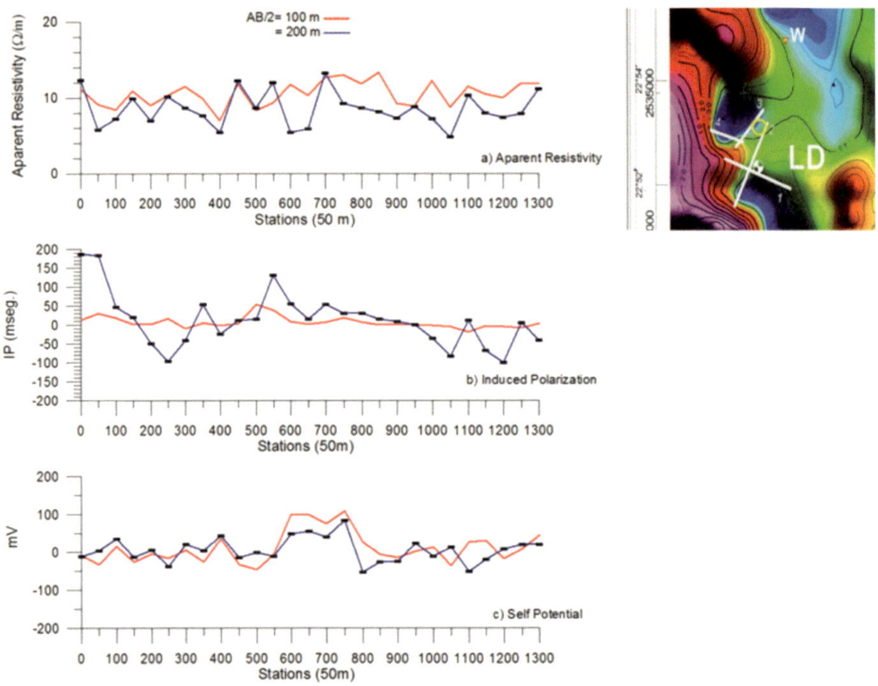

Figure 17.
Electrical section 2, with a NE–SW orientation, where (a) the apparent resistivity is plotted; in (b) the induced polarization and in (c) the self-potential. It is located on the magnetic section 2.

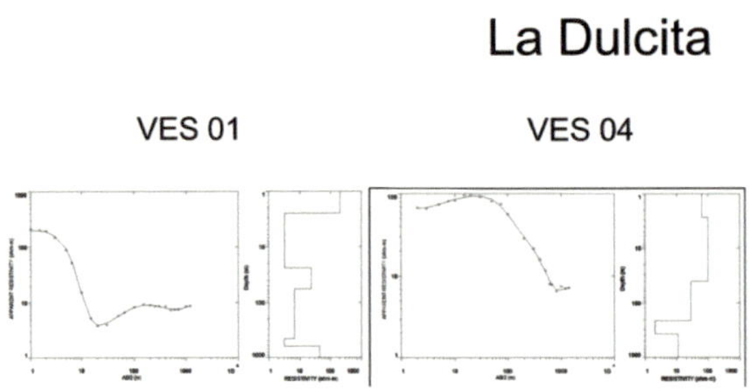

C#	RESISTIVITY	THICKNESS
	(ohm-m)	(meters)
1	205.7	2.47
2	3.21	21.25
3	23.23	33.54
4	6.68	398.9
5	3.06	163.0
6	43.66	

C#	RESISTIVITY	THICKNESS
	(ohm-m)	(meters)
1	64.78	2.70
2	101.1	37.47
3	28.69	170.4
4	2.02	157.8
5	10.84	

C#	RESISTIVITY	THICKNESS
	(ohm-m)	(meters)
1	102.4	2.44
2	18.65	30.25
3	9.11	167.6
4	6.32	674.8
5	81.10	

Figure 18.
Graphs of vertical electric soundings (VES) 1 and 2 and their comparison with VES 4, related to a well with an expenditure of the order of 25 L/s. note the thickness (170.4 m) correlated with the aquifer horizon (28.7 Ωm) possibly due to a sandy unit, overlying a clay horizon (2 Ωm).

permeability. One of the SEVs was carried out on a producer well that was located 2.3 km SW of La Dulcita and geologically located in the zone of the sunken block and aeromagnetically associated with the AMD I, which served as calibrator for the interpretations.

C#	RESISTIVITY	THICKNESS	C#	RESISTIVITY	THICKNESS	C#	RESISTIVITY	THICKNESS
	(ohm-m)	(meters)		(ohm-m)	(meters)		(ohm-m)	(meters)
1	53.78	3.35	1	64.78	2.70	1	684.6	4.50
2	4.45	58.09	2	101.1	37.47	2	48.89	11.11
3	53.20	34.22	3	28.69	170.4	3	14.81	29.08
4	3.48	228.0	4	2.02	157.8	4	9.24	719.1
5	55.22		5	10.84		5	27.06	

Figure 19.
Graphs of the vertical electric soundings (VES) 3 and 5 and its comparison with the VES 4, related to a well, with an yield of the order of 16 L/s. Note that VES 3 shows a sequence of geological units (~ 53 Ωm) confined by clay horizons (3–4 Ωm), while VES 5 shows a decrease in resistivity to depths of the order of 700 m.

The qualitative interpretation of the VES' morphology showed that the producer was well associated with a KQH curve (VES 4), among four remaining SEVs two were from the QQH family (VES 2 and 5), one was HKQH (VES 1) and the other HKH (VES 3).

The VES' were processed and interpreted with the commercial program Resix Plus that solves the inverse problem based on the Ghosh method of the inverse filter [22]. Each of the VES was compared with the VES (4) of the producing well (**Figures 18** and **19**).

Figure 18 indicates that the data interpreted in the VES 4 (KQH) for producing well and calibrator clearly indicates that at the base of the aquifer there is a clay unit (2 Ωm) and correlates with the resistivity of 28.69 Ωm with a thickness of 170.4 m, hence the well produces about 16 L/s. In this comparison, the VES 1 (HKQH) shows a horizon (23.23 Ωm) with a thickness of 33.5 m, possibly associating a sandy unit with moisture content at a depth of the order of 24 m. The VES 2 (QQH) shows a unit with a resistivity of 18.65 Ωm at a depth of less than 3 m with a thickness of 30 m.

Figure 19 shows the results of interpreting the VES' 3 and 5, they are also compared with the VES 4 (well). The VES 3 (HKH) shows the existence of a geological unit (53.20 Ωm) bordered by two clay horizons (4 and 3 Ωm) at a depth of the order of 61 m and a thickness of 34 m. It presents very good resistive contrast and the unit can be a fractured basalt horizon. VES 5 (QQH) shows a horizon possibly associated with a clay-sandy unit (14.8 Ωm) at a depth of 15 m and a thickness of 29 m. A large layer (> 700 m) of clay (9.2 Ωm) that starts to make an interpretation at a depth of 45 m.

5. Results and conclusions

Once the information was interpreted and analyzed, in some areas and communal land holding close to La Dulcita, a zone that meets the standards that are associated to the aquifer were found. It aeromagnetically shows the existence of alignments in N-S and E-W orientation and their location on top of the graben

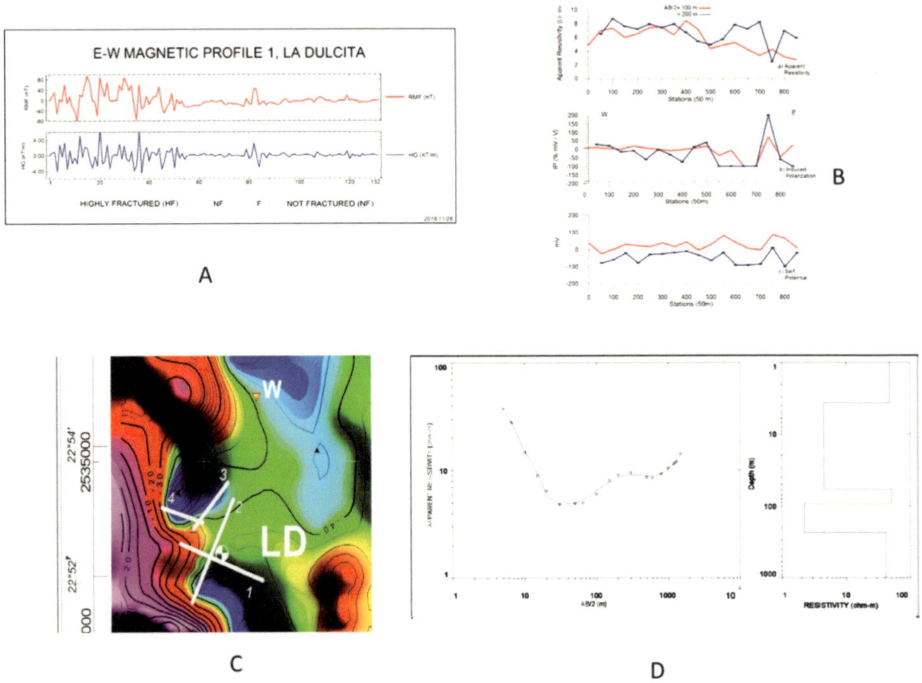

Figure 20.
The graphs that support the existence of a fractured area and with humidity in the underground are shown. The magnetism intensity graph (a) shows a clearly fractured zone towards the NW portion of the section. In the geoelectric profile (B), a contrast in resistivity is observed towards station 400, it decreases to openings of AB/2 = 200 m with respect to AB/2 = 100 m. the SEV 3 (C) shows association with type H curves.

structure. It is represented in an aeromagnetic map by the magnetic lows (blue color) the pit area and magnetic highs (color red). Thus, La Dulcita area is located in the limits of three aeromagnetic domains, which already indicates in ground magnetic measurements and should necessarily have a magnetic susceptibility contrast that will be reflected with significant differences in the amplitude of the magnetic field.

The ground magnetic sections indicate the zones that can be associated with permeability and the zones that do not have association with this physical property. The magnet simile is parameter for the interpretation of fracture in the underground. It will generate a simple anomaly if not related with fracture and provide a series of anomalies, which will be characterized by high frequencies. The calculation of the horizontal gradient of the magnetic field is completely resolutive to be able to observe fractured (permeable) zones of relatively healthy zones. In the magnetic section 1, the different physical behaviors that exist in the underground are clearly shown in the first portion of a highly fractured area contrasted with the rest of the section, which indicates that the magnetic susceptibilities of each terrestrial magnetic domains are associated with different units.

With the aerial and terrestrial magnetism, it was easy to find areas with high possibilities of being associated with fracturing (permeability).

With the electrical sections, it was possible to quickly scan the areas with possibilities of being associated with permeability and verify if they could also be associated with humidity. Producer well is key to facilitating the interpretation of vertical electric soundings, which in order to be associated with humidity should have as part of their morphology a portion type H.

In the area that was most likely to be associated with permeability and humidity in the underground (**Figure 20**) where a drilling was carried out by the State Water

Commission of San Luis Potosí, with production of 4 L/s. Furthermore, if we take into account the previously three dry wells, which had been drilled then it can be said that this methodology has met the objective.

With the help of the above methodology, the trained eye of the field geologist is strengthened with this methodology that uses scientific instruments, whose function is to detect the variation in the physical properties. For example, the magnetic susceptibility and resistivity of the rocks that are hidden below the Surface. Undoubtedly, the usage will certainly increase the percentage of successful drilled wells.

Acknowledgements

This work was funded by the State Water Commission, San Luis Potosí and COPOCYT-SLP. My sincere gratitude goes to Ing. Víctor J. Martínez Ruíz for his support to drawing geological map. I also thank David E. Torres Gaytán for his contribution in the preparation of this work. Also my sincere gratitude to Dr., Sanjeet Lumar Verna and Lucia Aldana Navarro for their comments on writing.

Author details

Héctor López Loera
Instituto Potosino de Investigación Científica y Tecnológica A.C. (IPICYT), Applied Geosciences Department, San Luis Potosí, México

*Address all correspondence to: hlopezl@ipicyt.edu.mx

References

[1] Keller GV, Frischknecht FC. Electrical Methods in Geophysical Prospecting. New York: Pergamos Press; 1966. 317 p

[2] Orellana E. Prospección Geoeléctrica en Corriente Continúa. Paraninfo: Biblioteca Técnica Philips; 1972. 523 p

[3] Kirch R. Groundwater Geophysics: A Tool for Hydrogeology. 2nd Edition: Springer; 2009. 548 p

[4] INEGI. 2010. Available at: http://www.inegi.org.mx/Sistemas/bise/mexicocifras/default.aspx?ent=24. (Accessed: July 2010)

[5] Freeze RA, Witherspoon PA. Theoretical analysis of regional groundwater flow, part 2. Effect of water-table configuration and subsurface permeability variations. Water Resources Research. 1967;**3**:623-634

[6] Tindall JA, Kunkel JR. Unsaturated Zone Hydrology for Environmental Scientists and Engineers. Boston MA: Prentice Hall; 1999. 625 p

[7] Labarthe HG, Aguillón RA. Cartografía Geológica 1:50,000, Hojas: Salinas y Villa de Ramos, Estados. De San Luis Potosí y Zacatecas. Folleto Técnico 106. Universidad Autónoma de San Luis Potosí, Instituto de Geología;1986. 52 p

[8] NOAA. 2010. Abilaable http://www.ngdc.noaa.gov/IAGA/vmod/igrf.html,

[9] Urrutia-Fucugauchi J, Campos-Enríquez JO. Geomagnetic secular variation in Central Mexico since 1923 AD and comparison with 1943-1990 IGRF models. Journal of Geomagnetism and Geoelectricity. 1993;**45**:243-249

[10] Baranov V, Naudy H. Numerical calculation of the formula of reduction to the magnetic pole. Geophysics. 1964;**29**:67-79

[11] Henderson RG, Zietz I. The computation of second vertical derivatives of geomagnetic fields. Geophysics. 1949;**14**:508-516

[12] Henderson RG. On the validity of the use of upward continuation integral for total magnetic intensity data. Geophysics. 1970;**35**:916-919

[13] Henkel H, Guzmán M. Magnetic features of fractures zones. Geoexploration. 1977;**5**:173-181

[14] López-Loer H, Urrutia-Fucugauchi J. Geophysical Study of Faulting Associated with the Colima Volcanic Complex: Volcan de Colima, Fifth International Meeting: Colima, Mexico, January 22-26, 1966, Abstract Volume. Colima: University of Colima; 1996

[15] Babu HVR, Rao NK, Kumar VV. Bedrock toporaphy from magnetic anomalies - an aid for groundwater exploration in hard- rocks terrains. Geophysics. 1991;**56**(07):1051-1054

[16] López-LoeraH,Urrutia-FucugauchiJ, Alva Valdivia LM. Magnetic characteristic of fracture zones and constraints on the subsurface structure of the Colima volcanic complex, western México. Geosphere. 2010;**6**(1):35-46

[17] McElhinny MW. Paleomagnetism and Plate Tectonics. London: Cambridge Earth Science Series; 1973. 358 p

[18] Tarling DH. Paleomagnetism. Principles and Applications in Geology, Geophysics and Archaeology. Chapman and Hall Ltd; 1983. 379 p

[19] Urrutia-Fucugauchi J. Importancia del magnetismo remanente natural

en la interpretación de las anomalías magnéticas. Boletín Asociación Mexicana de Geofísicos De Exploración. 1977;**18**(4):83-116

[20] Iakubovskii IUV, Liajov LL. Exploración Eléctrica. Moscú: Editorial Reverté, S.A.; 1980. 421 p

[21] Orellana E, Mooney HM. Master Tables and Curves for Vertical Electrical Soundings. España: Interciencia; 1966. 125 p

[22] Ghosh DP. Inverse filter coefficients for the computation of apparent resistivity standard curves for a horizontal stratified earth. Geophysical Prospecting. 1971;**19**(4):769-775

Section 3

Applications

Chapter 4

Atomic Scale Magnetic Sensing and Imaging Based on Diamond NV Centers

Myeongwon Lee, Jungbae Yoon and Donghun Lee

Abstract

The development of magnetic sensors simultaneously satisfying high magnetic sensitivity and high spatial resolution becomes more important in a wide range of fields including solid-state physics and life science. The nitrogen-vacancy (NV) center in diamond is a promising candidate to realize nanometer-scale magnetometry due to its excellent spin coherence properties, magnetic field sensitivity, atomic-scale size and versatile operation condition. Recent experiments successfully demonstrate the use of NV center in various sensing and imaging applications. In this chapter, we review the basic sensing mechanisms of the NV center and introduce imaging applications based on scanning magnetometry and wide field-of-view optics.

Keywords: magnetic sensor, diamond NV center, quantum sensing, scanning magnetometry, wide field-of-view optics

1. Introduction

Understanding magnetic properties at the microscopic level plays an important role in the development of modern science and technology [1, 2]. For instance, writing and reading information using nanometer size magnetic bits is the heart of massive data storage indispensable in the modern information technology [1]. Magnetic resonance imaging (MRI) which is an important medical tool of imaging the inner structures of human body is also based on sensing the magnetic response of minuscule protons with respect to radio frequency (RF) electromagnetic waves [2]. For fundamental research, on the other hand, studying magnetic phases and spin textures at the nanometer scale are one of the hottest topics in solid-state physics due to the recent discovery of exotic materials and topological phases [3–5]. Therefore, it is not too much to say that the continuous advances in modern science and technology strongly reply on the precise sensing and control of magnetism at the atomic level.

The paradigm of modern science and technology seems to shift from charge-based devices to spin-based systems. Nonetheless studying spins is a lot more difficult than electric charges mainly due to the lack of sensitive measurement techniques of magnetic field. For instance, the size of magnetic bits used in MOSFET (Metal-Oxide-Semiconductor Field-Effect Transistor) and STT-MRAM (Spin-Transfer Torque Magnetic Random-Access Memory) are less than 10 nanometers and eventually reaches at the level of single spins requiring sensitive detection of individual spins with high spatial resolution [6, 7]. However, detecting single electron spin takes more than 13 hours even with the best magnetometer [8], while sensing single electron charge takes only 1 picosecond [9, 10]. This motivates to develop new magnetic sensors with high magnetic field sensitivity and high spatial resolution.

Existing magnetometers are insufficient to satisfy both requirements especially when trying to measure minute magnetic fields at the length scale of 100 nm or below. For instance, SQUID (Superconducting Quantum Interference Devices), atomic vapor cell and Hall bar are very sensitive magnetometers but their spatial resolutions are typically limited to tens of micrometers [11, 12]. On the other hand, scanning probe type tools such as SP-STM (Spin-Polarized Scanning Tunneling Microscope) and MFM (Magnetic Force Microscope) exhibit very high spatial resolution but their sensitivity is relatively low and not quantitatively defined [13, 14]. Moreover, magnetic films coated at the tips may produce unwanted stray field affecting the magnetic samples to be measured.

Here, we introduce a novel magnetometer enabling non-invasive, extremely sensitive magnetic sensing and imaging at the nanometer scale. It is based on diamond NV (nitrogen-vacancy) center which is an atomic size point defect in the diamond crystal providing high spatial resolution. It is also a spin qubit (i.e. quantum bit) possessing remarkable magnetic and quantum properties satisfying high field sensitivity [15, 16]. Since it can also operate over a wide range of temperature from room temperature down to cryogenic temperatures and is chemically inert and non-toxic, the NV center already has been applied in various experiments including magnetic imaging of solid-state materials and biomedical samples [17, 18]. In this chapter, we will discuss basic working principles of diamond NV centers (Section 2) and their sensing mechanisms (Section 3). Furthermore, we will provide two examples of imaging applications; scanning probe type imaging (Section 4.1) and wide field-of-view optical imaging (Section 4.2) (**Figure 1**).

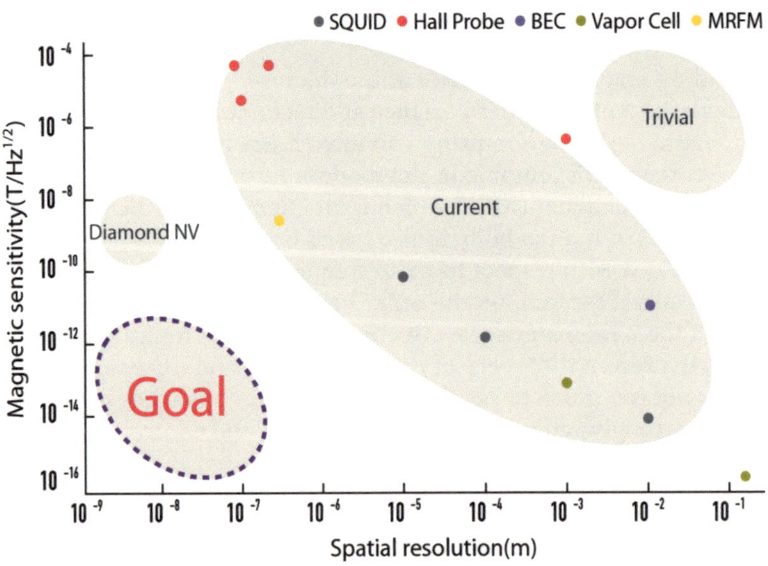

Figure 1.
Comparison of various magnetometers in terms of spatial resolution and magnetic field sensitivity. The diamond NV center is a promising candidate to realize the goal of highly sensitive sensing with nanometer-scale resolution. The data are adopted from [19]; MRFM [8]; SQUID [20–22]; Hall probe [23, 24]; BEC [25]; Vapor Cell [26].

2. Background of diamond NV center

The diamond NV center is a hetero-molecular defect in a diamond crystal consisting of a substitutional nitrogen defect combined with an adjacent carbon vacancy [15, 16] (**Figure 2a**). It is a color center as it absorbs photons in the visible range of wavelength (e.g. 532 nm) and emits photons of a broad range of wavelength (e.g. 632–800 nm). The NV center can exist in a natural diamond, but it can be created in more controllable fashion, for instance, by implanting nitrogen ions into the diamond. Subsequent high temperature annealing (e.g. at 800°C) results in the thermal migration of carbon vacancies and NV centers are formed once the vacancies meet the implanted nitrogen impurities. The density and location of NV centers in diamond are well controlled with various techniques. **Figure 2b** shows an example of precise positioning of NV centers less than a few hundred of nanometers uncertainty [27].

When negatively charged, the NV center has total six electrons (i.e. three electrons from three carbons, two electrons from the substitutional nitrogen and one electron from the diamond lattice). Four of them form pairs and the remaining two unpaired electrons make spin triplet states (i.e. S = 1) in the ground energy level. The spin triplet states are split into $m_s = 0$ and $m_s = \pm 1$ whose separation is about 2.9 GHz at room temperature due to the crystal field and spin–spin interaction [17, 18] (this is called zero-field splitting). As shown in **Figure 3a**, the degenerated $m_s = \pm 1$ states are split further if there is non-zero magnetic field along the NV crystal axis (i.e. the quantized axis of NV spin). Sensing magnetic field (i.e. magnetic field component along the NV axis) is realized by measuring the amount of Zeeman splitting [17, 18] (e.g. 5.6 MHz splitting per 1 Gauss field).

The optical response of the NV center varies depending on its spin states. When it is in the $m_s = 0$ state, almost 100% cycling transition occurs upon optical pumping. On the other hand, for the case of $m_s = \pm 1$ states, 10–30% of the excited electrons undergo intersystem crossing (ISC) to the spin singlet states and relax into the $m_s = 0$ ground state. This dark transition results in the reduction of the number of emitted photons relative to the $m_s = 0$ state. Furthermore, the transition via the shelving states produces spin flip from the $m_s = \pm 1$ states to the $m_s = 0$ state. The spin-sensitive fluorescence and spin-flip transition allow optical readout of the spin states as well as optical initialization of the qubit state [15–18].

The energy levels of NV center are located well within the bandgap of diamond (i.e. 5.3 eV) making it effectively isolated from the hosting material and enabling to preserve its intrinsic quantum properties. Thus, the NV center has exceptionally long spin coherence times even at room temperature [28] (e.g. $T_2 > 1$ ms). In

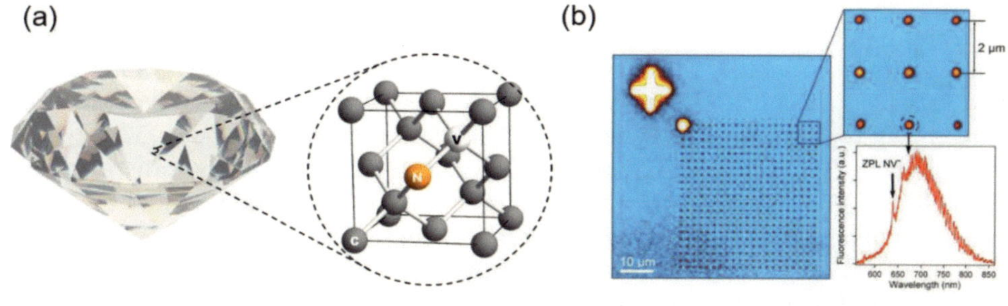

Figure 2.
Physical properties of diamond NV center. (a) Crystal structure of NV center in a diamond lattice. NV center consists of nitrogen substitutional defect and carbon vacancy. (b) Precise formation of NV centers using focused ion beam (FIB) implantation of nitrogen. Reprint with permission from [27]. Copyright (2013) Wiley-VCH Verlag GmbH & Co. KGaA. Reproduced with permission.

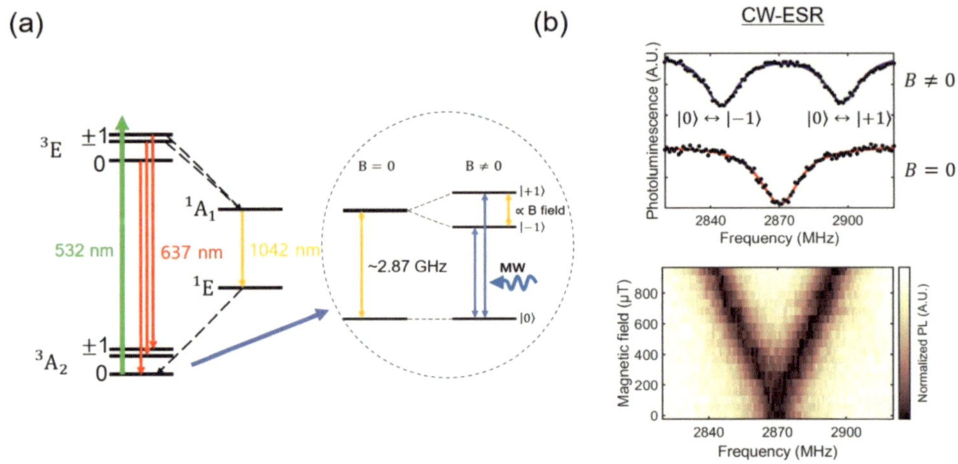

Figure 3.
Electronic properties of diamond NV center. (a) Energy levels and optical transitions are illustrated. The spin triplet ground states are used to realize a spin qubit for the magnetic field sensing. (b) Photoluminescence signal as a function of microwave frequency reveals the Zeeman splitting providing information about the external magnetic field. This method is called CW-ESR measurement.

addition, the NV spin is highly sensitive to various fields, including temperature, magnetic, electric and strain fields [18]. For example, the magnetic field sensitivity for a single NV center is on the order of 1 $nT/\sqrt{Hz}$ which is about 10^5 smaller than typical earth magnetic field [16–18]. Higher sensitivity can be also possible using either NV ensembles [29] (e.g. $<1\ pT/\sqrt{Hz}$) or advanced sensing protocols [30]. The high magnetic field sensitivity is one of the key ingredients of realizing novel magnetometer introduced in this chapter.

3. Sensing mechanism

In this section, we will discuss various sensing methods specifically designed to study magnetism in different spectral regimes from static spin distributions to high frequency magnetic excitations. Sections 3.1 and 3.2 elucidate basic mechanisms used for sensing static and dynamic magnetic fields. The section ends with a brief outlook on advanced sensing techniques (Section 3.3).

3.1 Sensing dc magnetic field

Probing static field from magnetic textures or current flow in transport devices is an important capability to study microscopic magnetism in condensed matter physics. The diamond NV centers already have been used to study a diverse set of magnetic systems including skyrmions [31], superconducting vortices [32, 33], domain walls [34], magnetic nanowires [35], and steady-state current distributions in graphene [36]. Various experimental methods have been implemented to detect static dc magnetic field and we will examine the two most common protocols; continuous wave electron spin resonance (CW-ESR) and Ramsey interferometry.

As seen in **Figure 3b**, the ground spin states of NV center are subject to change by external magnetic field via the Zeeman effect. Upon continuous illumination of the pumping laser and gigahertz microwave photons, ESR transitions of $m_s = 0 \leftrightarrow m_s = -1$ and $m_s = 0 \leftrightarrow m_s = +1$ occurs. The ESR signals appear as negative peaks in the photoluminescence (PL) measurement due to the dark transitions associated with the $m_s = \pm 1$ states. Since the ESR spectrum is obtained by optical means, this method is often called as optically detected magnetic resonance (ODMR).

The amount of dc magnetic field is extracted from the ESR splitting of $\Delta f = 2\gamma B_{NV}$, where Δf is the frequency difference of the two transitions, γ is the gyromagnetic ratio (i.e. 2.8 MHz/G) and B_{NV} is the magnetic field along the NV axis. The CW-ESR signal can be analyzed by fitting the spectrum with a Lorentzian peak,

$$I(f) = I_0 \left(1 - C \frac{(f_{FWHM}/2)^2}{(f - f_0)^2 + (f_{FWHM}/2)^2} \right) \tag{1}$$

where I_0 is the photon count rate, C is the optical contrast between the spin states, f_0 is the ESR central frequency, and f_{FWHM} is the full width half maximum of the resonance. The sensitivity of dc magnetic field based on this method is determined by the minimum resolvable frequency shift (**Figure 4**) which is defined as $\Delta f = \frac{\Delta N}{\partial N/\partial f}$, where N is total number of photons during the measurement time, T, i.e. $N = I(f)T$. For the shot noise limited photon measurement, $\Delta N \approx \sqrt{N}$ and the minimum detectable magnetic field and the sensitivity become [16].

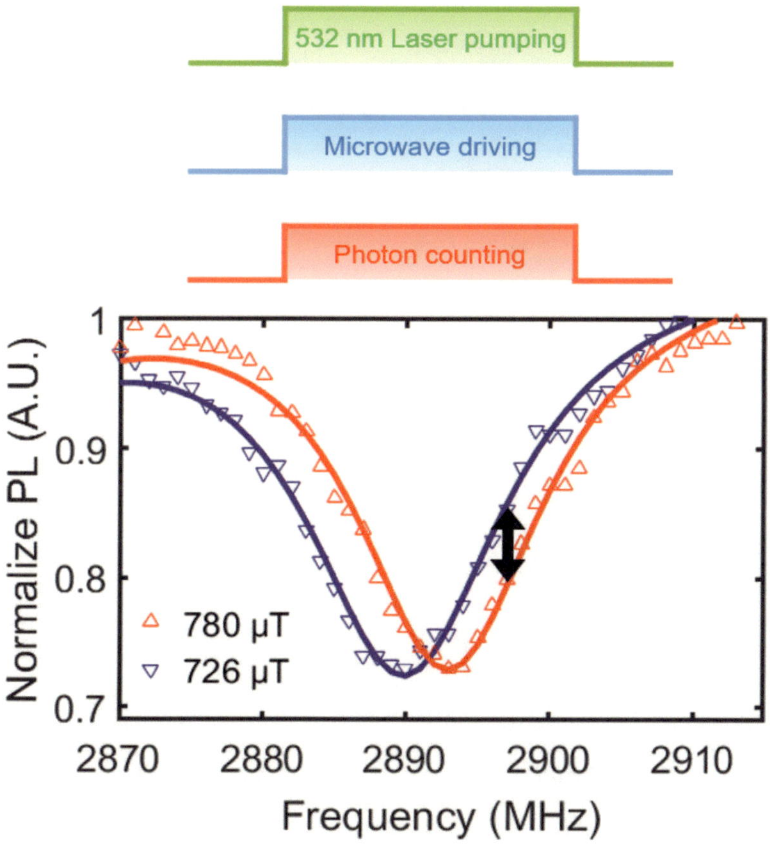

Figure 4.
Sensing dc magnetic field based on CW-ESR measurement. Continuous wave of laser and microwave photons are used. The shift in frequency gives an information about the magnitude and direction of external magnetic field.

$$B_{min}=\frac{\Delta f}{\gamma}\approx\frac{\sqrt{I_0 T}}{\gamma C I_0 T/f_{\text{FWHM}}}=\frac{f_{\text{FWHM}}}{\gamma C\sqrt{I_0 T}},\quad \eta_B=B_{min}\sqrt{T}\approx\frac{f_{\text{FWHM}}}{\gamma C\sqrt{I_0}}. \tag{2}$$

As seen in Eq. (2), the sensitivity is limited by the ESR linewidth and in principle it can be as narrow as the inverse of NV's inhomogeneous dephasing time, T_2^*. However, practical linewidth suffers from the power broadening due to continuous laser and microwave excitation. Therefore, typical magnetic sensitivity based on CW-ESR is limited to on the order of $1\,\mu T/\sqrt{\text{Hz}}$.

The power broadening problem can be avoided by using pulsed laser and microwave photons. Ramsey interferometry is one of the basic pulse techniques used to measure the free induction decay of a spin qubit. **Figure 5a** shows Ramsey pulse sequences used in the NV measurement. The two laser pulses are used for optical initialization and readout while the two $\pi/2$ microwave pulses are used to make qubit superposition state and to project it back to the initial state. The basic idea of Ramsey interferometry is very similar with Michelson interferometry. The first laser pulse polarizes the NV center to the $m_s = 0$ state and the subsequent $\pi/2$ microwave pulse rotates the spin into the equal superposition of $m_s = 0$ and $m_s = +1$ (or $m_s = -1$) state. This works as a 50:50 beam splitter used in the Michelson interferometry experiment. Under external magnetic field, the two spin states evolve together but with different phases each other and the amount of accumulated phase depends on the magnitude of dc field. If the microwave frequency is detuned from the qubit energy by δ, the Ramsey signal oscillates at the frequency of δ and is written as

$$I(t)\approx I_0\left(1-\frac{C}{2}\right)+I_0\frac{C}{2}\left(e^{-\left(\frac{t}{T_2^*}\right)^2}\cos\delta t\right) \tag{3}$$

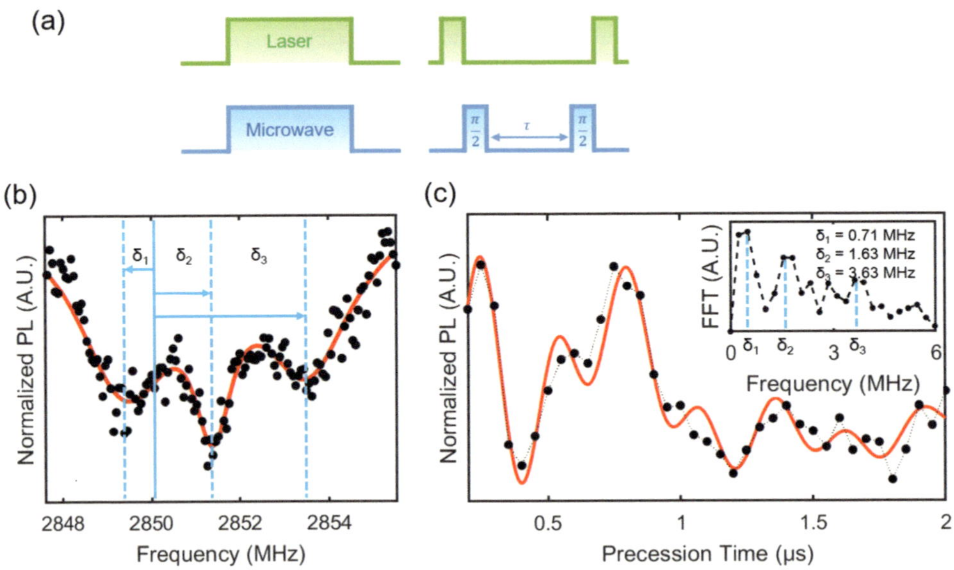

Figure 5.
Sensing dc magnetic field based on Ramsey interferometry. (a) Schematics of laser and microwave pulse used for the measurement. (b) A close view of the resonance peak in CW-ESR reveals three sub-features resulting from the NV's hyperfine coupling with ^{14}N nuclear spin. (c) Ramsey measurement with the detunings in (b) shows the beatings of three oscillations. Subset presents the FFT of the Ramsey signal showing three peaks corresponding to the detunings.

where t is the free induction time. Shift in the oscillation frequency gives the information about static magnetic field and the resolvable frequency shift is now limited by T_2^*. The minimum detectable magnetic field and the sensitivity are written as.

$$B_{min} = \frac{\Delta\delta}{\gamma} \approx \frac{\sqrt{I_0 t_r n}}{\gamma C I_0 t_r n t} e^{\left(\frac{t}{T_2^*}\right)^2}, \eta_B \approx B_{min}\sqrt{n(t_r + t)} \approx B_{min}\sqrt{nt} = \frac{1}{\gamma C \sqrt{I_0 t_r}\sqrt{T_2^*}} \approx \frac{1}{\sqrt{T_2^*}} \quad (4)$$

where t_r is the single shot readout time (e.g. a few hundreds of nanoseconds) and n is the total number of measurement cycles [16]. Based on this method, the dc field sensitivity from a single NV center can be as high as $\eta_B \sim 10\ \mathrm{nT}/\sqrt{\mathrm{Hz}}$ for $T_2^* \sim 100\ \mu s$ [16].

Figure 5b and **c** shows an example of the Ramsey measurement. The ESR resonance exhibits three hyperfine structures due to the dipole-dipole interaction between NV electron spin and ^{14}N nuclear spin (I = 1). Therefore, the beatings of three oscillations appear in the Ramsey signal (**Figure 5c**) which is defined as,

$$I(t) = I_0\left(1 - \frac{C}{2}\right) + I_0\frac{C}{2}\left(e^{-\left(\frac{t}{T_2^*}\right)^2}\frac{1}{3}\sum_i^3 \cos\left(\delta_i t + \phi_i\right)\right) \quad (5)$$

where δ_i $(i = 1, 2, 3)$ are the detunings of three hyperfine levels and ϕ_i are the phase offsets. The fast Fourier transformation (FFT) of the signal also reveals three frequencies (subset in **Figure 5c**) whose shifts are used to probe static field.

3.2 Sensing ac magnetic field

The NV center can also detect ac magnetic field up to gigahertz. The large bandwidth sensing capability is important to study spin dynamics in solid-state systems and biological samples. For example, nuclear spin precession in biomolecules occurs at 100 kHz–MHz regime while spin excitations or charge fluctuations in solids happen at higher frequency of MHz–GHz [30]. Various sensing methods with its detection bandwidth are listed in **Figure 6** and, in this section, we will

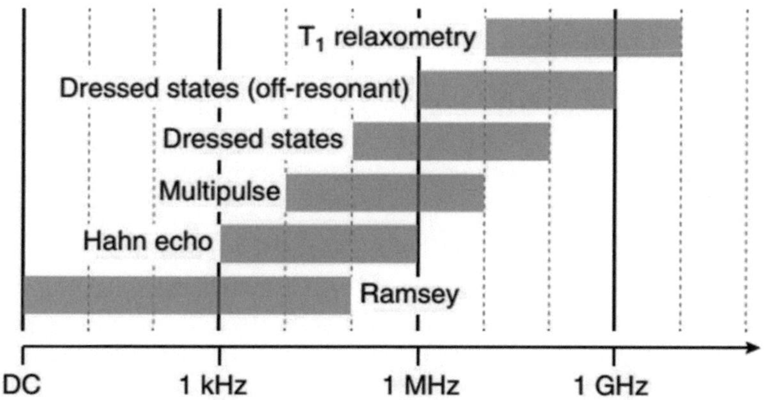

Figure 6.
Detection bandwidth of various sensing methods. NV center can measure magnetic field from dc to gigahertz ac frequency. Reprint with permission from Ref. [30]. Copyright (2017) Reviews of Modern Physics.

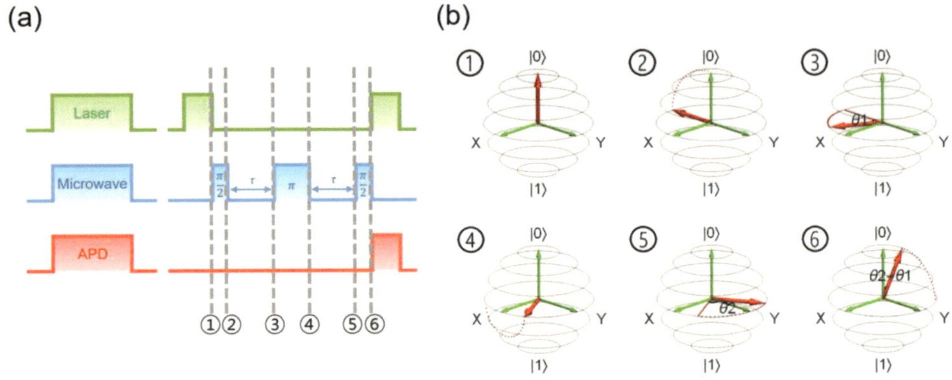

Figure 7.
Sensing ac magnetic field based on spin Hahn echo measurement. (a) Schematics of the Hahn echo pulses. (b) Bloch representations of the spin status according to the numbers indicated in (a).

explain two basic dynamical decoupling methods such as spin Hahn echo and Carr-Purcell-Meiboom-Gill (CPMG).

Spin Hahn echo measurement consists of a similar pulse sequence as Ramsey interferometry but having an extra π pulse in the middle of the sequence (**Figure 7a**). This additional pulse flips the sign of accumulated phase during the free induction evolution resulting in the cancelation of dc and low frequency magnetic field. When the pulse duration matches to the period of ac magnetic field, however, the phase survives and continues to be accumulated. In this way, one can selectively probe ac magnetic field.

This can be better viewed with the Bloch sphere representation shown in **Figure 7b**. In the rotating frame at the qubit frequency, the spin rotates around the equator by an angle, θ, which is determined by the free precession time, τ, between $\pi/2$ and π pulses (② and ⑤ in **Figure 7b**). The angles from the first (②) and the second period (⑤) of the echo pulse are written as,

$$\theta_1 = \int_0^{\tau} \gamma B_{AC} \sin(\omega t' + \alpha) dt', \theta_2 = \int_{\tau}^{2\tau} \gamma B_{AC} \sin(\omega t' + \alpha) dt' \quad (6)$$

where B_{AC} is the amplitude of ac magnetic field and α is a phase offset. And their difference is

$$\theta_2 - \theta_1 = \frac{4\gamma B_{AC}}{\omega} \cos(\omega\tau + \alpha) \sin^2\frac{\omega\tau}{2} \quad (7)$$

This difference determines the projected probability of the qubit and becomes non-zero if the sign of B_{AC} flips before and after the π pulse. The probabilities of the $|0\rangle$ and $|1\rangle$ states after a single echo sequence are obtained as,

$$P_{|1\rangle} = \sin^2\left(\frac{\theta_2 - \theta_1}{2}\right), P_{|0\rangle} = \cos^2\left(\frac{\theta_2 - \theta_1}{2}\right). \quad (8)$$

Finally, an average probability of the qubit state after repeated Hahn echo cycles (but with random phase offset, α) can be expressed as the first order Bessel function,

$$P_{|0\rangle, avg} = \frac{1}{2\pi} \int_{-\pi}^{\pi} \cos^2\left[\frac{2\gamma B_{AC}}{\omega} \cos(\omega\tau + \alpha) \sin^2\frac{\omega\tau}{2}\right] d\alpha = \frac{1}{2}\left[1 + J_0\left(\frac{4\gamma B_{AC}}{\omega} \sin^2\frac{\omega\tau}{2}\right)\right]. \quad (9)$$

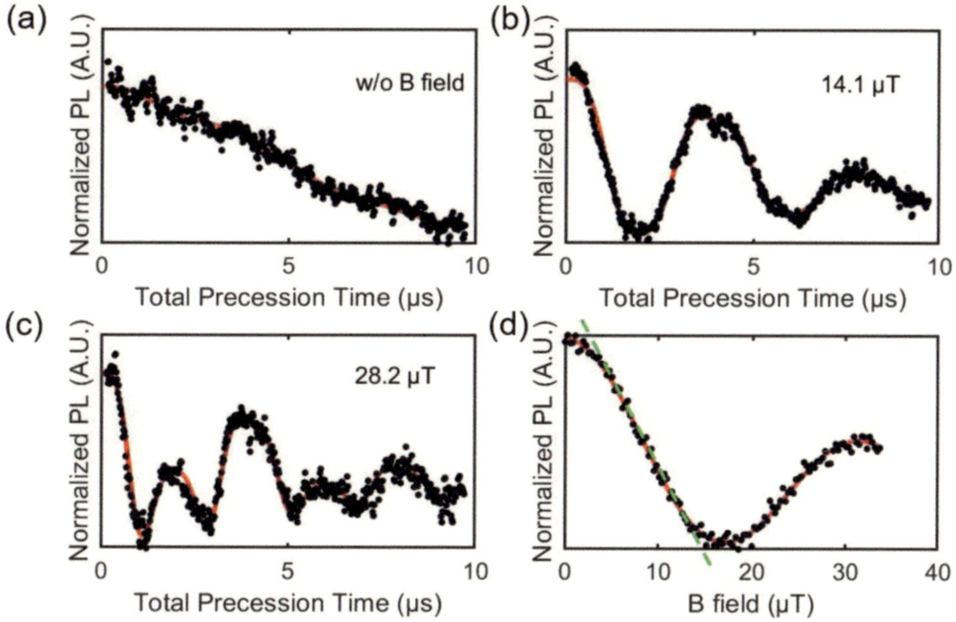

Figure 8.
Examples of the Hahn echo measurement as a function of ac field strength. (a–c) Hahn echo signals with various strengths of ac magnetic field at 500 kHz. (a) no field, (b) 14.1 μT, and (c) 28.2 μT. (d) Hahn echo signals as a function of the magnetic field strength measured at $2\tau = 2\,\mu s$. Minimum detectable magnetic field of $B_{min} = 0.84 \pm 0.02\,\mu$T is obtained from the dashed line of the maximum slope.

Figure 8 shows an example of the spin Hahn echo measurement as a function of the free precession time, τ, and the ac field strength, B_{AC}. For the experiment, external field at 500 kHz frequency is applied with a wire. Compared to a monotonic exponential decay in **Figure 8a**, the first order Bessel functions discussed in Eq. (9) are clearly observed for the cases of non-zero ac fields (**Figure 8b** and **c**). The field sensitivity can be determined from the maximum change of the PL signal with respect to B_{AC}. For instance, **Figure 8d** shows the normalized PL as a function of B_{AC} measured at the time, $2\tau = 2\,\mu s$. The minimum detectable field (or sensitivity) is obtained from the ratio of the maximum slope over the noise level in the PL signal (shot noise limited in this measurement) which is $B_{min} = 0.84 \pm 0.02\,\mu$T at 500 kHz.

By adding more periodic microwave pulses in the sequence, one can extend the spin coherence time and realize improved sensitivity. For example, CPMG sequence utilizes n pairs of two π pulses separated by 2τ (**Figure 9**) which prolong the spin coherence by $T_2 \propto n^{2/3}$. The axis of π pulse can be also alternated between x and y in the Bloch sphere which can mitigate potential pulse error. Such dynamical decoupling sequences are called XY4 or XY8.

The ac field sensitivity is similar as the dc field sensitivity in Eq. (4) but now the intrinsic spin dephasing time, T_2, limits the sensitivity i.e. $\eta_{B,AC} \approx \eta_{B,DC}\sqrt{\frac{T_2^*}{T_2}}$. The ac field sensitivity based on as a single NV center can be as high as $\eta_B \sim 1\,\text{nT}/\sqrt{\text{Hz}}$ for $T_2 \sim 1$ ms [16].

3.3 Current limitations and advanced sensing protocols

In general, the bandwidth of dynamical decoupling methods is limited to ~ 10 MHz (**Figure 6**). Higher frequency ac field (~GHz) can be measured by T_1

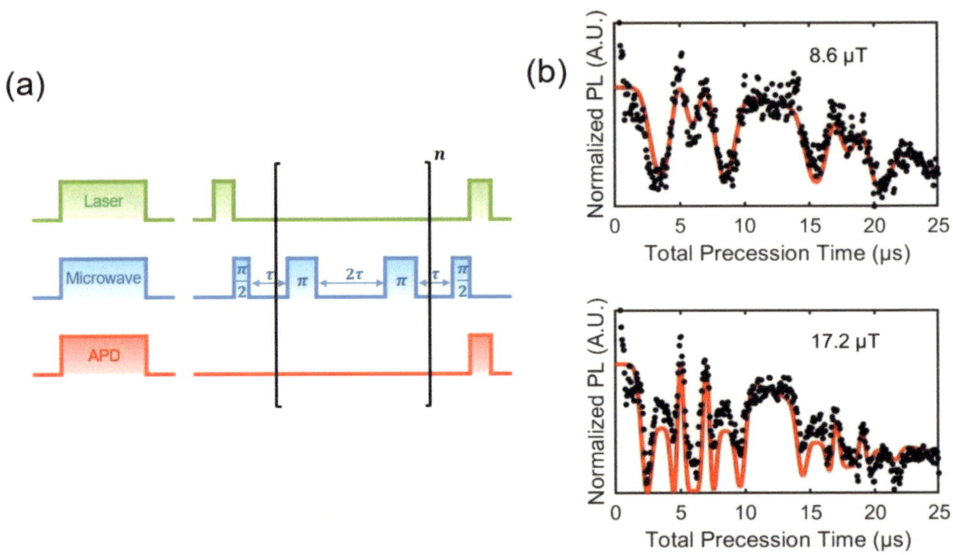

Figure 9.
Sensing ac magnetic field based on CPMG measurement. (a) Schematics of the CPMG pulses. (b) CPMG signals with two different ac magnetic fields at 500 kHz, i.e. 8.6 and 17.2 μT.

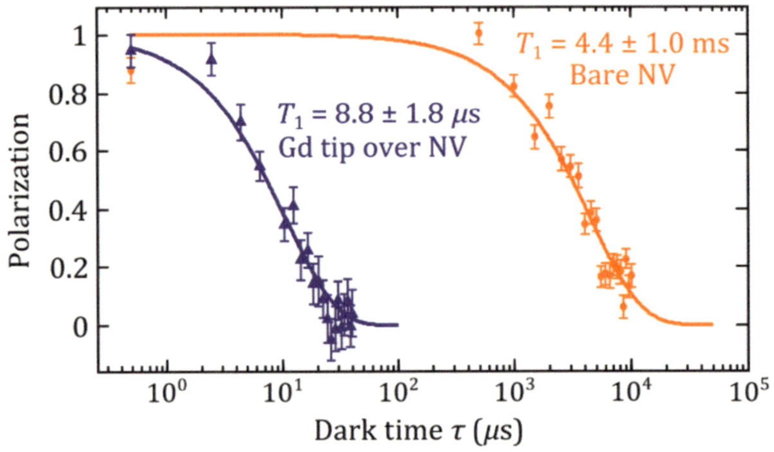

Figure 10.
An example of T_1 relaxometry. Fast fluctuation of Gd^{3+} spins results in the reduction of NV's T_1 time enabling sensing of gigahertz frequency ac field. Reprint with permission from Ref. [37]. Copyright (2014) Physical Review Applied.

relaxometry [18, 30]. Dynamic field fluctuation around the qubit frequency can directly affect the qubit relaxation time which can be measured by the relaxometry technique. For instance, **Figure 10** shows that gigahertz spin fluctuations of Gd^{3+} ions result in much faster relaxation of the qubit [37].

In terms of the field sensitivity, higher sensitivity can be realized using either ensembles of NV center or advanced sensing protocols. The former is possible since the sensitivity is proportional to $\sqrt{N}$ where N is the number of NV centers. For instance, sub-picotesla sensitivity has been demonstrated based on an ensemble of $N \sim 10^{11}$ NV centers [29]. The latter relays on advanced sensing methods utilizing various quantum techniques such as sensing assisted by entanglement or auxiliary qubits [30]. For example, nuclear spins typically exhibit 1000 times longer

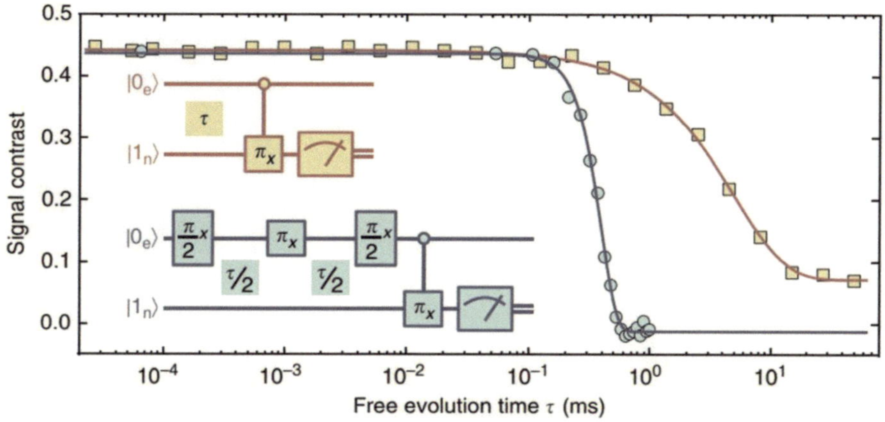

Figure 11.
An example of quantum memory measurement. Spin coherence time is extended by using a neighboring nuclear spin as a memory qubit. Reprint with permission from [38]. Copyright (2016) Nature Communications.

coherence time compared to electron spins such NV center. By using a nuclear spin as an auxiliary qubit via entanglement with the NV electron spin, one can realize extended coherence time and obtain enhanced sensitivity. This method is called quantum memory and an example is shown in **Figure 11** [38].

4. Imaging applications

In this section, we will introduce magnetic imaging techniques based on the diamond NV center. The combined properties of atomic-scale size and high field sensitivity make the NV center as a novel imaging tool to study nanoscale magnetism in the field of condensed matter physics and biology. Multiple imaging techniques have been implemented in several experiments and we will discuss two examples of the diamond imaging techniques; scanning magnetometry and wide field-of-view optics.

4.1 Scanning magnetometry for solid-state systems

Scanning probe microscopy (SPM) is a versatile tool to image sample surface with high spatial resolution. The probe tip can scan over the surface with nanometer step size while maintaining vertical distance with feedback techniques. Scanning tunneling microscope (STM) and atomic force microscope (AFM) are the most common SPM methods used in various experiments. In NV-based imaging applications, different geometries of SPM are possible depending on the position of NV center either on tip or on surface. For example, magnetic molecules that are attached at the end of AFM tip are scanned over a bulk diamond surface where NV centers are located underneath the surface [37] (e.g. 5–20 nm). On the other hand, the NV center can be positioned at the apex of AFM tip and is scanned over magnetic samples [31–34]. There are several benefits of the former geometry. A simple SPM design is possible and there are no needs for complicated diamond fabrication. Moreover, one can use NV centers in a bulk diamond which typically possess good coherence properties. However, this method is not suitable to image large area of sample since it is not easy to be prepared on the tip. Therefore, the geometry of NV-on-tip is more suitable to study solid-state materials.

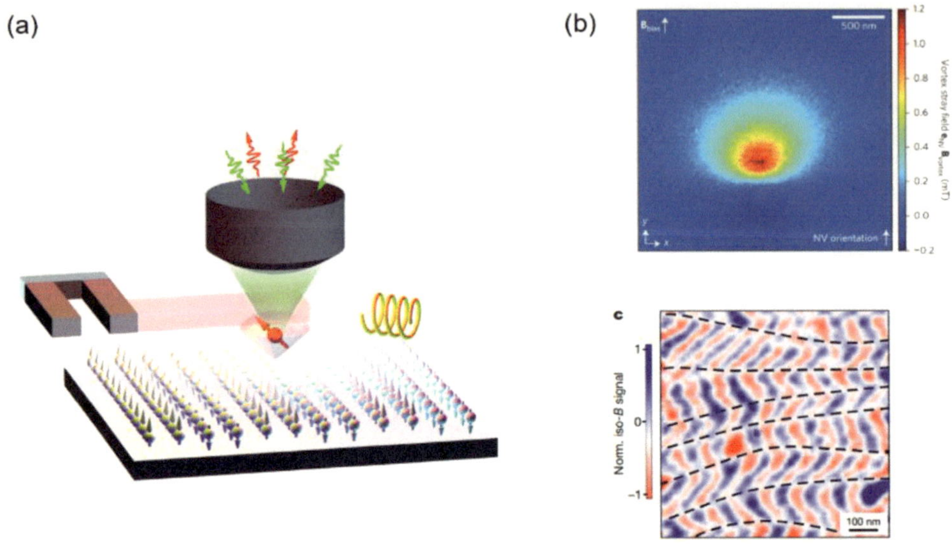

Figure 12.
Scanning magnetometry based on diamond NV center. (a) A schematic of diamond scanning magnetometer. A diamond probe scans over a magnetic sample while an objective lens collects NV photons and a wire produces microwave excitations. (b) Examples of magnetic imaging on various magnetic samples such as superconducting vortex, multiferroic magnetic orders (following from the top image to the bottom image). Reprint with permissions from [39] Copyright (2017) Nature, [32] Copyright (2017) Nature Nanotechnology.

Figure 12a shows a schematic of the NV-on-tip geometry. A fabricated diamond probe with pillar structures or a diamond nano-particle is glued at the end of an AFM tuning fork. An objective lens focuses on the NV center at the tip apex for the optical excitation and readout. A scanning stage maneuvers the sample in three dimensional directions with nanometer step size. In this way, the NV center can effectively scan over the sample and detects local magnetic fields at every scan position on the surface. **Figure 12b** shows examples of the scan images on various magnetic samples including superconducting vortex [32, 33], and multiferroic materials [39]. Since the NV center can operate from room temperature down to cryogenic temperatures, this novel method provides an efficient way to study temperature-dependent evolution of magnetic orders in exotic materials.

4.2 Wide field-of-view optics for life science

The diamond SPM method provides high resolution imaging but is quite slow due to the scanning process (e.g. a few hours per image) and is not an efficient method to study dynamical features. Fast magnetic imaging can be realized by simultaneous mapping the sample with wide field-of-view optics. While confocal optics used in the SPM collects photons only from a focused spot, wide field-of-view optics records optical signals from every point within the optical field-of-view (e.g. $100 \times 100\ \mu m$) with a CCD (Charge-coupled device) camera. Even though the spatial resolution is not as good as the SPM methods and is diffraction limited, the faster imaging capability gives more advantages when one studies biological samples.

In the diamond wide field-of-view optics experiment, ensembles of NV center are typically used to enhance the field sensitivity and biological samples are placed on a bulk diamond containing NV ensembles. Simultaneous excitation of a number group of NV centers within the field-of-view requires high power of pumping laser. In order to avoid potential damage to the bio-sample due to the high power laser

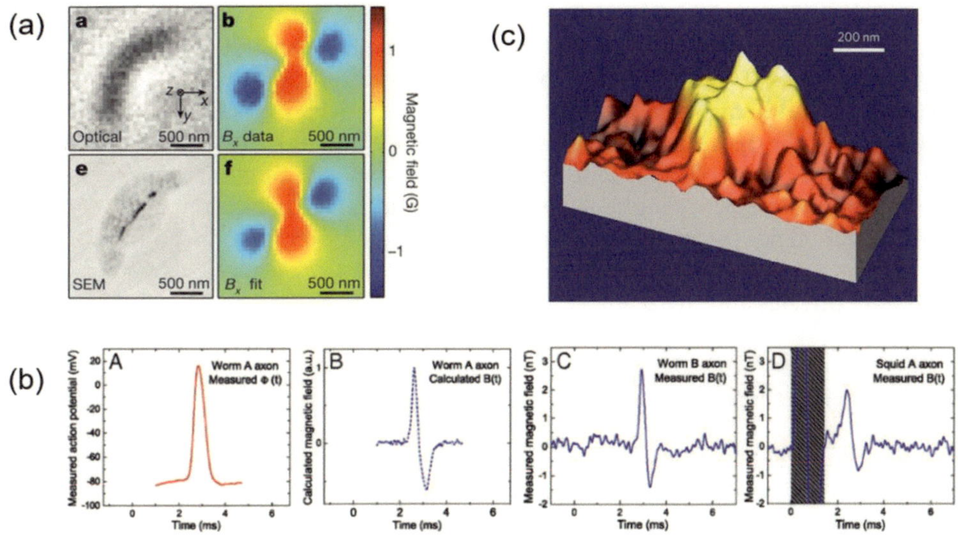

Figure 13.
Examples of magnetic sensing and imaging on biological samples. (a) Magnetic imaging on a single magnetotactic bacterium. Reprint with permission from [40] Copyright (2013) Nature (b) sensing of induced magnetic field due to the action potential in an axon. Reprint with permission from [41] Copyright (2016) Proceedings of the National Academy of Sciences (c) MRI imaging of a particle of poly(methyl methacrylate) (PMMA). Reprint with permission from [42] Copyright (2015) Nature Nanotechnology.

beam, total internal reflection fluorescence (TIRF) technique is typically adopted [40]. With the TIRF configuration, the excitation laser from the backside of the diamond is totally reflected at the interface between the diamond and the sample and is only illuminated onto the NV ensembles.

Figure 13a shows an example of the wide field-of-view magnetic imaging on biological samples e.g. bacteria called magnetotactic bacteria (MTB) [40]. MTB contain magnetic nano-particles in the body forming one dimensional chain that and are aligned along the external magnetic field. The diamond wide field-of-view optics successfully maps the stray field produced by a single MTB and identifies orientations of the magnetic chains from the field distribution.. Motivated by this work, scientists try to identify cancer cells among normal cells by selectively dosing magnetic nano-particles into the target cells and imaging the resulting magnetic field [43].

Owing to the capability of bio-imaging, the NV center gets increasing attentions in the field of neural science. Since neurons communicate each other via the motions of ions, it can be viewed as current flows in a conducting wire which produces magnetic field around it. By mapping the induced magnetic field with the NV center, therefore, one can study brain activities in the neural networks. As a first step toward this goal, recent experiment successfully demonstrates detection of magnetic field due to the action potential along an axon (**Figure 13b**) [41]. The NV center is also used to probe nuclear magnetic resonance (NMR) signal from a single protein [44] which opens up the possibilities of MRI at the nanometer-scale [42] (**Figure 13c**).

5. Conclusions

In this chapter, we introduce the diamond NV center as a novel magnetometer satisfying high magnetic field sensitivity and high spatial resolution. We review the basic working principles of sensing dc and ac magnetic field and discuss two imaging applications which are based on scanning magnetometry and wide field-of-view optics techniques. The excellent properties of NV centers such as sub-nanotesla sensitivity, nanometer-scale resolution, gigahertz range of detection bandwidth, wide range of the operation temperature, non-toxic and bio-friendly host material position the NV center as a unique tool of sensing and imaging microscopic magnetic phenomena. Improving sensing protocols and imaging techniques is on-going efforts in this field. In conclusion, the novel magnetometer introduced in this chapter has a promising potential to be used in various research fields particularly solid-state physics and life science.

Acknowledgements

The authors acknowledge the support from the Basic Science Research Program through the National Research Foundation of Korea (NRF) funded by the Ministry of Education (NRF-2016R1D1A1A02937119) and the Korea University Future Research Grant (K1822781).

Author details

Myeongwon Lee, Jungbae Yoon and Donghun Lee*
Department of Physics, Korea University, Seoul, Republic of Korea

*Address all correspondence to: donghun@korea.ac.kr

References

[1] Soaldin NA. Magnetic materials. 2nd ed. Fundamentals and Applications. Cambridge: Cambridge University Press; 2010. pp. 177–188. ISBN-13: 978–0521886697

[2] Shenton ME, Hamoda HM, Schneiderman JS, Bouix S, Pasternak O, Rathi Y, et al. A review of magnetic resonance imaging and diffusion tensor imaging findings in mild traumatic brain injury. Brain Imaging and Behavior. 2012;**6**(2):137-192. DOI: 10.1007/s11682-012-9156-5

[3] Hsieh D, Qian D, Wray L, Xia Y, Hor YS, Cava RJ, et al. A topological Dirac insulator in a quantum spin hall phase. Nature. 2008;**452**:970-974. DOI: 10.1038/nature06843

[4] Eerenstein W, Mathur ND, Scott JF. Multiferroic and magnetoelectric materials. Nature. 2006;**442**:759-765. DOI: 10.1038/nature05023

[5] Nagaosa N, Tokura Y. Topological properties and dynamics of magnetic skyrmions. Nature Nanotechnology. 2013;**8**:899-911. DOI: 10.1038/NNANO.2013.243

[6] Kuhn K, Kenyon C, Kornfeld A, Liu M, Maheshwari A, Shih W, et al. Managing process variation in Intel's 45 nm CMOS technology. Intel Technology Journal. 2008;**12**:93-109. DOI: 10.1535/itj.1202

[7] Parkin S, Jiang X, Kaiser C, Panchula A, Roche K, Samant M. Magnetically engineered spintronic sensors and memory. Proceedings of the IEEE. 2003; **91**:661-680. DOI: 10.1109/JPROC.2003.811807

[8] Rugar D, Budakian R, Hamin HJ, Chui BW. Single spin detection by magnetic resonance force microscopy. Nature. 2004;**430**:329-332. DOI: 10.1038/nature02658

[9] Schoelkopf RJ, Wahlgren P, Kozhevnikov AA, Delsing P, Prober DE. The radio-frequency single-electron transistor (RF-SET): A fast and ultrasensitive electrometer. Science. 1998;**280**:1238-1242. DOI: 10.1126/science.280.5367.1238

[10] Aassime A, Johansson G, Wendin G, Schoelkopf RJ, Delsing P. Radio-frequency single-electron transistor as readout device for Qubits: Charge sensitivity and Backaction. Physical Review Letters. 2001;**86**:3376-3379. DOI: 10.1103/PhysRevLett.86.3376

[11] Hasselbach K, Veauvy C, Mailly D. MicroSQUID magnetometry and magnetic imaging. Physica C. 2000;**332**: 140-147. DOI: 10.1016/S0921-4534(99)00657-7

[12] Oral A, Bending SJ. Real-time scanning Hall probe microscopy. Applied Physics Letters. 1996;**69**: 1324-1326. DOI: 10.1063/117582

[13] Wiesendanger R. Spin mapping at the nanoscale and atomic scale. Reviews of Modern Physics. 2009;**81**:1495-1550. DOI: 10.1103/RevModPhys.81.1495

[14] Hartmann U. Magnetic force microscopy. Annual Review of Materials Science. 1999;**29**:53-87. DOI: 10.1146/annurev.matsci.29.1.53

[15] Childress L, Hanson R. Diamond NV centers for quantum computing and quantum networks. MRS Bulletin. 2013; **38**:134-138. DOI: 10.1557/mrs.2013.20

[16] Taylor JM, Cappellaro P, Childress L, Jiang L, Budker D, Hemmer PR, et al. High-sensitivity diamond magnetometer with nanoscale resolution. Nature Physics. 2008;**4**: 810-816. DOI: 10.1038/nphys1075

[17] Rondin L, Tetienne JP, Rohart S, Thiaville A, Hingant T, Spinicelli P, et al.

Stray-field imaging of magnetic vortices with a single diamond spin. Nature Communications. 2013;**4**:1-5. DOI: 10.1038/ncomms3279

[18] Schrhagl R, Chang K, Loretz M, Degen CL. Nitrogen-vacancy centers in diamond: Nanoscale sensors for physics and biology. Annual Review of Physical Chemistry. 2014;**65**:83-105. DOI: 10.1146/annurev-physchem-040513-103659

[19] Maze JR. Quantum Manipulation of Nitrogen-Vacancy Centers in Diamond: From Basic Properties to Applications. Harvard University; 2010. Available from ProQuest Dissertations & Thesis Global: https://search.proquest.com/docview/612776202?accountid=14558 [Accessed: 2019-02-01]

[20] Vengalattore M, Higbie JM, Leslie SR, Guzman J, Sadler LE, Stamper-Kurn DM. High-Resolution Magnetometry with a Spinor Bose-Einstein Condensate. Physical Review Letters. 2017;**98**: 200801. DOI: 10.1103/PhysRevLett.98.200801

[21] Faley MI, Poppe U, Urban K, Paulson DN, Fagaly RL. A new generation of the HTS multilayer DC-SQUID magnetometers and gradiometers. Journal of Physics: Conference Series. 2006;**43**:1199-1202. DOI:: 10.1088/1742-6596/43/1/292

[22] Baudenbacher F, Fong LE, Holzer JR, Radparvar M. Monolithic low-transition-temperature superconducting magnetometers for high resolution imaging magnetic fields of room temperature samples. Applied Physics Letters. 2003;**82**:3487-3489. DOI: 10.1063/1.1572968

[23] SANDHU A, Okamoto A, Shibasaki I, Oral A. Nano and micro Hall-effect sensors for room-temperature scanning hall probe microscopy. Microelectronic Engineering. 2004;**73-74**:524-528. DOI: 10.1016/j.mee.2004.03.029

[24] Sandhu A, Kurosawa K, Dede M, Oral A. Japanese Journal of Applied Physics. 2004;**43**:777-778. DOI: 10.1143/JJAP.43.777

[25] Vengalattore M, Higbie JM, Leslie SR, Guzman J, Sadler LE, Stamper-Kurn DM. High-Resolution Magnetometry with a Spinor Bose-Einstein Condensate. Physical Review Letters. 2007;**98**: 200801. DOI: 10.1103/PhysRevLett.98.200801

[26] Shah V, Knappe S, Schwindt PDD, Kitching J. Subpicotesla atomic magnetometry with a microfabricated vapour cell. Nature Photonics. 2007;**1**: 649-652. DOI: 10.1038/nphoton.2007.201

[27] Lesik M, Spinicelli P, Pezzagna S, Happel P, Jacques V, Salord O, et al. Maskless and targeted creation of arrays of colour centres in diamond using focused ion beam technology. Physica Status Solidi (a). 2013;**210**:2055-2059. DOI: 10.1002/pssa.201300102

[28] Bar-Gill N, Pham LM, Jarmola A, Budker D, Walsworth RL. Solid-state electronic spin coherence time approaching one second. Nature Communications. 2013;**4**:1743. DOI: 10.1038/ncomms2771

[29] Wolf T, Neumann P, Nakamura K, Sumiya H, Ohshima T, Isoya J, et al. Subpicotesla diamond magnetometry. Physical Review X. 2015;**5**:041001. DOI: 10.1103/PhysRevX.5.041001

[30] Degen CL, Reinhard F, Cappellaro P. Quantum sensing. Reviews of Modern Physics. 2017;**89**:035002. DOI: 10.1103/RevModPhys.89.035002

[31] Dovzhenko Y, Casola F, Schlotter S, Zhou TX, Büttner F, Walsworth RL, et al. Magnetostatic twists in room-temperature skyrmions explored by nitrogen-vacancy center spin texture reconstruction. Nature

Communications. 2018;**9**:2712. DOI: 10.1038/s41467-018-05158-9

[32] Thiel L, Rohner D, Ganzhorn M, Appel P, Neu E, Müller B, et al. Quantitative nanoscale vortex imaging using a cryogenic quantum magnetometer. Nature Nanotechnology. 2016;**11**:677-681. DOI: 10.1038/nnano.2016.63

[33] Pelliccione M, Jenkins A, Ovartchaiyapong P, Reetz C, Emmanouilidou E, Ni N, et al. Scanned probe imaging of nanoscale magnetism at cryogenic temperatures with a single-spin quantum sensor. Nature Nanotechnology. 2016;**11**:700-705. DOI: 10.1038/nnano.2016.68

[34] Tetienne JP, Hingant T, Kim JV, Diez LH, Adam JP, Garcia K, et al. Nanoscale imaging and control of domain-wall hopping with a nitrogen-vacancy center microscope. Science. 2014;**344**:1366-1369. DOI: 10.1126/science.1250113

[35] Lee M, Jang B, Yoon J, Mathpal MC, Lee Y, Kim C, et al. Magnetic imaging of a single ferromagnetic nanowire using diamond atomic sensors. Nanotechnology. 2018;**29**:405502. DOI: 10.1088/1361-6528/aad2fe

[36] Tetienne JP, Dontschuk N, Broadway DA, Stacey A, Simpson DA, Hollenberg LCL. Quantum imaging of current flow in graphene. Science Advances. 2017;**3**:e1602429. DOI: 10.1126/sciadv.1602429

[37] Pelliccione M, Myers BA, Pascal LMA, Das A, Bleszynski Jayich AC. Two-dimensional nanoscale imaging of gadolinium spins via scanning probe relaxometry with a single spin in diamond. Physical Review Applied. 2014;**2**:054014. DOI: 10.1103/PhysRevApplied.2.054014

[38] Zaiser S, Rendler T, Jakobi I, Wolf T, Lee S-Y, Wagner S, et al. Enhancing quantum sensing sensitivity by a quantum memory. Nature Communications. 2016;**7**:12279. DOI: 10.1038/ncomms12279

[39] Gross I, Akhtar W, Garcia V, Martínez LJ, Chouaieb S, Garcia K, et al. Real-space imaging of non-collinear antiferromagnetic order with a single-spin magnetometer. Nature. 2017;**549**: 252-256. DOI: 10.1038/nature23656

[40] Le Sage D, Arai K, Glenn DR, DeVience SJ, Pham LM, Rahn-Lee L, et al. Optical magnetic imaging of living cells. Nature. 2014;**496**:486-489. DOI: 10.1038/nature12072

[41] Barry JF, Turner MJ, Schloss JM, Glenn DR, Song Y, Lukin MD, et al. Optical magnetic detection of single-neuron action potentials using quantum defects in diamond. Proceedings of the National Academy of Sciences. 2016;**113**: 14133-14138. DOI: 10.1073/pnas.1601513113

[42] Rugar D, Mamin HJ, Sherwood MH, Kim M, Rettner CT, Ohno K, et al. Proton magnetic resonance imaging using a nitrogen-vacancy spin sensor. Nature Nanotechnology. 2015;**10**: 120-124. DOI: 10.1038/nnano.2014.288

[43] Glenn DR, Lee K, Park H, Weissleder R, Yacoby A, Lukin MD, et al. Single-cell magnetic imaging using a quantum diamond microscope. Nature Methods. 2015;**12**:1-5. DOI: 10.1038/nmeth.3449

[44] Lovchinsky I, Sushkov AO, Urbach E, de Leon NP, Choi S, De Greve K, et al. Nuclear magnetic resonance detection and spectroscopy of single proteins using quantum logic. Science. 2016;**351**: 836-841. DOI: 10.1126/science.aad8022

Chapter 5

SQUID Magnetometers, Josephson Junctions, Confinement and BCS Theory of Superconductivity

Navin Khaneja

Abstract

A *s*uperconducting *qu*antum *i*nterference *d*evice (SQUID) is the most sensitive magnetic flux sensor currently known. The SQUID can be seen as a flux to voltage converter, and it can generally be used to sense any quantity that can be transduced into a magnetic flux, such as electrical current, voltage, position, etc. The extreme sensitivity of the SQUID is utilized in many different fields of applications, including biomagnetism, materials science, metrology, astronomy and geophysics. The heart of a squid magnetometer is a tunnel junction between two superconductors called a Josephson junction. Understanding the work of these devices rests fundamentally on the BCS theory of superconductivity. In this chapter, we introduce the notion of local potential and confinement in superconductivity. We show how BCS ground state is formed from interaction of wave packets confined to these local potential wells. The starting point of the BCS theory of superconductivity is a phonon-mediated second-order term that describes scattering of electron pair at Fermi surface with momentum $k_i, -k_i$ and energy $2\hbar\omega_i$ to $k_j, -k_j$ with energy $2\hbar\omega_j$. The transition amplitude is $\mathcal{M} = -\frac{d^2\omega_d}{(\omega_i-\omega_j)^2-\omega_d^2}$, where d is the phonon scattering rate and ω_d is the Debye frequency. However, in the presence of offset $\omega_i - \omega_j$, there is also a present transition between states $k_i, -k_i$ and $k_j, -k_i$ of sizable amplitude much larger than $\mathcal{M}$. How are we justified in neglecting this term and only retaining $\mathcal{M}$? In this chapter, we show all this is justified if we consider phonon-mediated transition between wave packets of finite width instead of electron waves. These wave packets are in their local potentials and interact with other wave packets in the same well to form a local BCS state we also call BCS molecule. Finally, we apply the formalism of superconductivity in finite size wave packets to high T_c in cuprates. The copper electrons in narrow d-band live as packets to minimize the repulsion energy. The phonon-mediated coupling between wave packets (of width Debye energy) is proportional to the number of k-states in a packet, which becomes large in narrow d-band (10 times s-band); hence, d-wave T_c is larger (10 times s-wave). At increased doping, packet size increases beyond the Debye energy, and phonon-mediated coupling develops a repulsive part, destroying superconductivity at large doping levels.

Keywords: local potentials, superconductivity, phonons, Josephson junctions, squids

1. Introduction

There is a very interesting phenomenon that takes place in solid-state physics when certain metals are cooled below critical temperature of order of few Kelvin. The resistance of these metals completely disappears and they become superconducting. How does this happen? One may guess that maybe at low temperatures there are no phonons. That is not true, as we have low frequency phonons present. Why do we then lose all resistivity? Electrons bind together to form a molecule by phonon-mediated interaction. The essence of this interaction is that electron can pull on the lattice which pulls on another electron. This phonon-mediated bond is not very strong for only few meV, but at low temperatures, this is good enough; we cannot break it with collisions with phonons which only carry $k_B T$ amount of energy which is small at low temperatures. Then, electrons do not travel alone; they travel in a bunch, as a big molecule; and you cannot scatter them with phonon collisions.

This phenomenon whereby many materials exhibit complete loss of electrical resistance when cooled below a characteristic critical temperature [1, 2] is called superconductivity. It was discovered in mercury by Dutch physicist Onnes in 1911. For decades, a fundamental understanding of this phenomenon eluded the many scientists who were working in the field. Then, in the 1950s and 1960s, a remarkably complete and satisfactory theoretical picture of the classic superconductors emerged in terms of the Bardeen-Cooper-Schrieffer (BCS) theory [3]. Before we talk about the BCS theory, let us introduce the notion of local potentials.

Shown in **Figure 1** is a bar of metal. How are electrons in this metal bar? Solid-state physics texts start by putting these electrons in a periodic potential [4–6]. But that is not the complete story.

Shown in **Figure 2** is a periodic array of metal ions. Periodic arrangement divides the region into cells (region bounded by dashed lines in **Figure 2**) such that the potential in the ith cell has the form

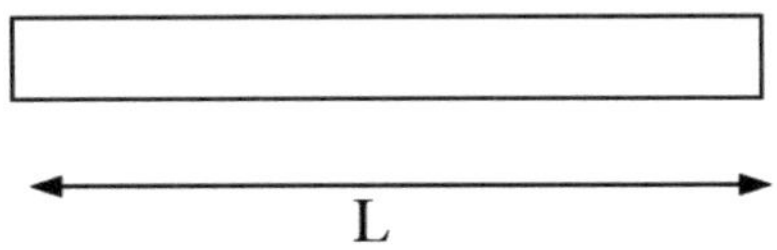

Figure 1.
Depiction of a metallic bar.

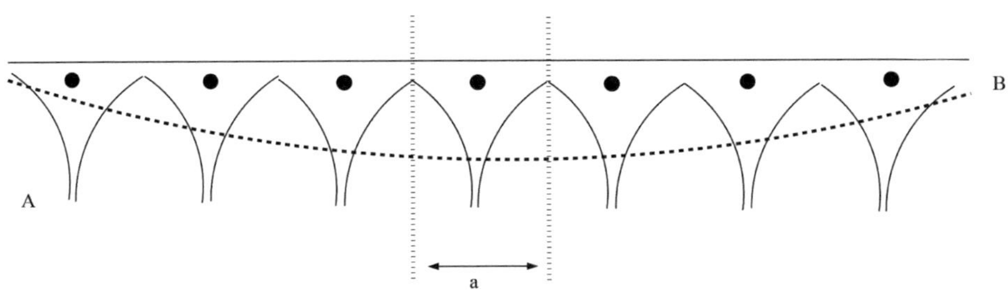

Figure 2.
Depiction of the potential due to metal ions. A rapidly varying periodic part A *and slowly varying part* B.

$$V(x) = \underbrace{\frac{-Ze}{4\pi\epsilon_0}\frac{1}{|x-a_i|}}_{A} + \underbrace{\frac{-Ze}{4\pi\epsilon_0}\sum_{j\neq i}\frac{1}{|x-a_j|}}_{B}, \tag{1}$$

where a_i is co-ordinate of the ion in the cell i. Eq. 1 has part A, the own ion potential. This gives the periodic part of the potential shown as thick curves in **Figure 2** and part B, the other ion potential. This is shown as dashed curve (trough or basin) in **Figure 2**. We do not talk much about this potential in solid-state physics texts though it is prominent (for infinite lattice, part B is constant that we can subtract). For $Z = 1$ and $a = 3A^\circ$, we have $V_0 = \frac{e}{4\pi\epsilon_0 a} \sim 3V$. Then, the other ion potential has magnitude $\ln n \;\; V_0$ at its ends and $2\ln\frac{n}{2} \;\; V_0$ in the centre. The centre is deeper by $\sim \ln\frac{n}{2}V_0$. Let us say metal block is of length 30 cm with total sites $n = 10^9$; then, we are talking about a trough that is ~50 V deep. At room temperature, the kinetic energy of the electrons in $\frac{1}{2}k_B T = .02eV$ (as we will see subsequently, Fermi-Dirac statistics give much higher kinetic energies), the trough is deep enough to confine these electrons. It is this trough, basin or confining potential that we talk about in this chapter. Electrons move around in this potential as wave packets as shown in **Figure 3A**. Electrons can then be treated as a Fermi gas as shown in **Figure 3B**. While gas molecules in a container rebound of wall, the confining potential ensures electrons roll back before reaching the ends.

Coming back to a more realistic estimate of the kinetic energy, electron wave function is confined to length $L = na$ (due to confining potential); then, we can expand its eigenfunctions $\exp\left(i\frac{2\pi m}{na}x\right) = \exp(ikx)$ with energies $\frac{\hbar^2 k^2}{2m}$. We fill each k state with two electrons with $\frac{-\pi}{2a} \leq k \leq \frac{\pi}{2a}$, and kinetic energy of electrons goes from 0 to $\frac{\hbar^2\pi^2}{2a^2 m} \sim 5$ eV. This spread of kinetic energy is modified in the presence of periodic potential. We then have an energy band as shown in **Figure 4** with a bandwidth of 5–10 eV.

With this energy bandwidth, electrons are all well confined by the confining potential. In fact we do not need a potential of depth 50 eV; to confine the electrons, we can just do it with a depth of ~10 eV which means a length of around $L_0 \sim 300$ Å. It means electrons over length L_0 are confined, and due to *screening* by electrons outside L_0, they simply do not see any potential from ions outside this length. Thus, we get a local confining potential, and the picture is shown in **Figure 5**, many local wells. This is what we call *local potentials or local volumes*. Estimate of L_0 is a 1D calculation; in 3D it comes to a well of smaller diameter. However, if we account for *electron–electron repulsion*, then our earlier 1D estimate is probably okay. In any case, these numbers should be taken with a grain of salt. They are more qualitative than quantitative. The different wave packets in a local volume do not leave the volume as

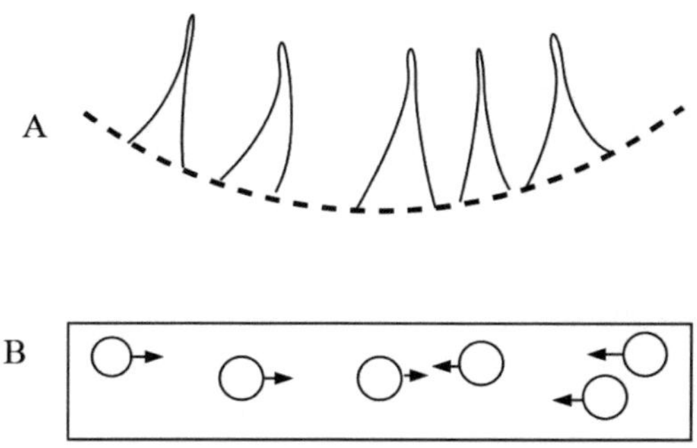

Figure 3.
Depiction of Fermi gas of electrons in a metal moving in a confining potential.

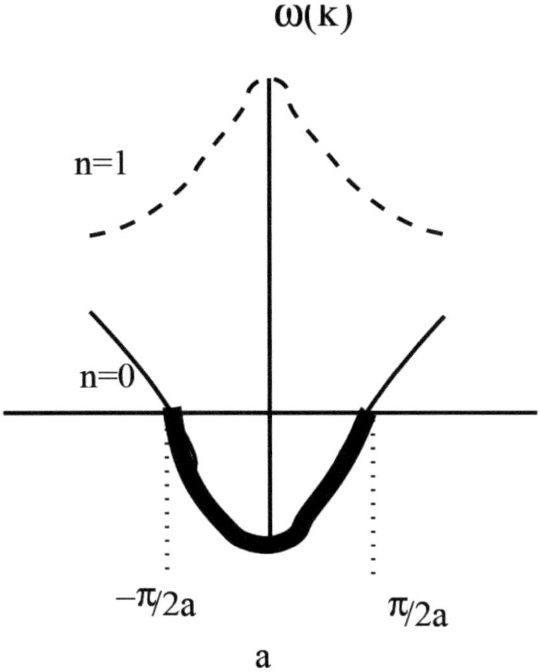

Figure 4.
The dispersion curve and energy band for electrons in a periodic potential.

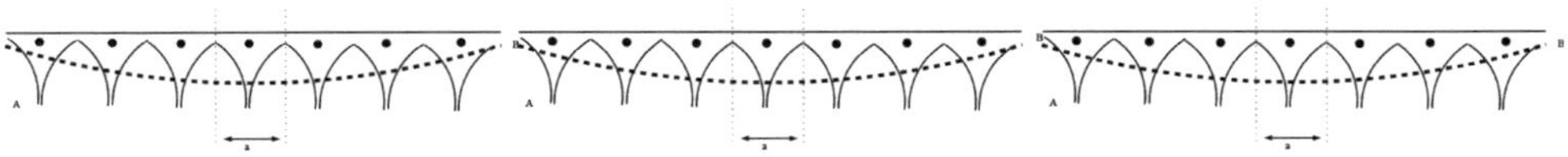

Figure 5.
Depiction of local potentials due to metal ions that locally confine electrons.

they see a local potential due to positive atomic ions. They just move back and forth in a local volume. When we apply an electric field say along x direction, the wave packets accelerate in that direction, and the local volume moves in that direction as a whole. This is electric current. Electrons are moving at very high velocity up to 10^5 m/s (Fermi velocity) in their local volumes, but that motion is just a back-and-forth motion and does not constitute current. The current arises when the local volume moves as a whole due to applied electric field. This is much slower at say drift velocity of 10^{-3} m/s for an ampere current through a wire of cross section 1 mm^2.

In this chapter, we spell out the main ideas of the BCS theory. The BCS theory tells us how to use phonon-mediated interaction to bind electrons together, so that we have big molecule and we call the BCS ground state or the BCS molecule. At low temperatures, phonons do not have energy to break the bonds in the molecule; hence, electrons in the molecule do not scatter phonons. So, let us see how BCS binds these electrons into something big.

2. Cooper pairs and binding

Let us take two electrons, both at the Fermi surface, one with momentum k_1 and other $-k_1$. Let us see how they interact with phonons. Electron k_1 pulls/plucks on the lattice due to Coulomb attraction and in the process creates (emits) a phonon and thereby recoils to new momentum k_2. The resulting lattice vibration is sensed

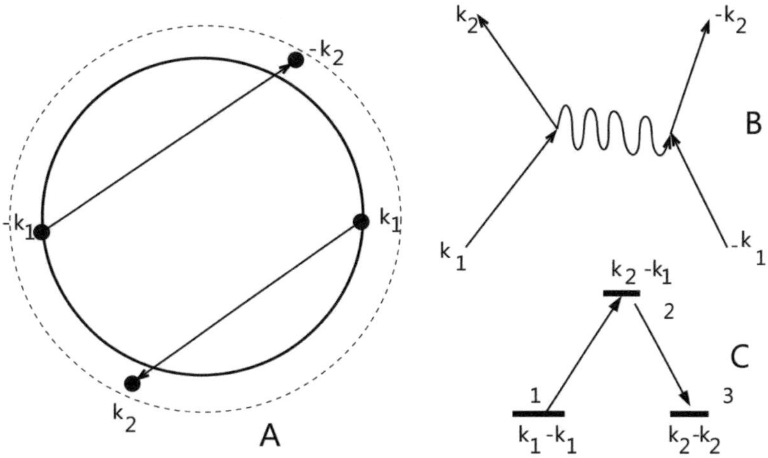

Figure 6.
(A) Depiction of the Fermi sphere and how electron pair $k_1, -k_1$ at Fermi sphere scatters to $k_2, -k_2$ at the Fermi sphere. (B) How this is mediated by exchange of a phonon in a Feynman diagram. (C) A three-level system that captures the various transitions involved in this process.

by electron $-k_1$ which absorbs this oscillation and is thrown back to momentum $-k_2$. The total momentum is conserved in the process. This is depicted in **Figure 6A**. The corresponding Feynman diagram for this process is shown in **Figure 6B**. The above process where two electrons interact with exchange of phonon can be represented as a three-level atomic system. Level 1 is the initial state of the electrons $k_1, -k_1$, level 3 is the final state of the electrons $k_2, -k_2$ and the level 2 is the intermediate state $k_2, -k_1$. There is a transition with strength $\Omega = \hbar d$ between levels 1 and 2 involving emission of a phonon and a transition with strength Ω between levels 2 and 3 involving absorption of a phonon. Let E_1, E_2, E_3 be the energy of the three levels. Since pairs are at Fermi surface, $E_1 = E_3$. The state of the three-level system evolves according to the Schrödinger equation:

$$\dot{\psi} = \frac{-i}{\hbar} \begin{bmatrix} E_1 & \Omega & 0 \\ \Omega^* & E_2 & \Omega^* \\ 0 & \Omega & E_1 \end{bmatrix} \psi. \tag{2}$$

We proceed into the interaction frame of the natural Hamiltonian (system energies) by transformation

$$\phi = \exp\left(\frac{it}{\hbar} \begin{bmatrix} E_1 & 0 & 0 \\ 0 & E_2 & 0 \\ 0 & 0 & E_1 \end{bmatrix}\right) \psi. \tag{3}$$

This gives for $\Delta E = E_2 - E_1$

$$\dot{\phi} = \frac{-i}{\hbar} \underbrace{\begin{bmatrix} 0 & \exp\left(-\frac{i}{\hbar}\Delta E\ t\right)\Omega & 0 \\ \exp\left(\frac{i}{\hbar}\Delta E\ t\right)\Omega^* & 0 & \exp\left(\frac{i}{\hbar}\Delta E\ t\right)\Omega^* \\ 0 & \exp\left(-\frac{i}{\hbar}\Delta E\ t\right)\Omega & 0 \end{bmatrix}}_{H(t)} \phi. \tag{4}$$

$H(t)$ is periodic with period $\Delta t = \frac{2\pi}{\Delta E}$. After Δt, the system evolution is

$$\phi(\Delta t) = \left(I + \int_0^{\Delta t} H(\sigma) d\sigma + \int_0^{\Delta t} \int_0^{\sigma_1} H(\sigma_1) H(\sigma_2) d\sigma_2 d\sigma_1 + \ldots \right) \phi(0). \tag{5}$$

The first integral averages to zero, while the second integral

$$\int_0^{\Delta t} \int_0^{\sigma_1} H(\sigma_1) H(\sigma_2) d\sigma_2 d\sigma_1 = \frac{1}{2} \int_0^{\Delta t} \int_0^{\sigma_1} [H(\sigma_1), H(\sigma_2)] d\sigma_2 d\sigma_1. \tag{6}$$

Evaluating it explicitly, we get for our system that second-order integral is

$$\frac{-i\Delta t}{\hbar} \begin{bmatrix} 0 & & \frac{|\Omega|^2}{E_1 - E_2} \\ 0 & 0 & 0 \\ \underbrace{\frac{|\Omega|^2}{E_1 - E_2}}_{\mathcal{M}} & 0 & 0 \end{bmatrix}, \tag{7}$$

which couples levels 1 and 3 and drives transition between them at rate $\mathcal{M} = \frac{\Omega^2}{E_1 - E_2}$.

Observe $E_1 = 2\epsilon_1$ and $E_3 = 2\epsilon_2$ and the electron energies $E_2 = \epsilon_1 + \epsilon_2 + \epsilon_d$, where $\epsilon_d = \hbar\omega_d$ is the energy of emitted phonon. We have $\epsilon_1 = \epsilon_2 = \epsilon$. Then, the transition rate between levels 1 and 3 is $\mathcal{M} = -\frac{\Omega^2}{\epsilon_d}$. Therefore, due to interaction mediated through lattice by exchange of phonons, the electron pair $k_1, -k_1$ scatters to $k_2, -k_2$ at rate $-\frac{\Omega^2}{\epsilon_d}$. The scattering rate is in fact $\Delta_b = -\frac{4\Omega^2}{\epsilon_d}$ as k_1 can emit to k_2 or $-k_2$. Similarly, $-k_1$ can emit to k_2 or $-k_2$, making it a total of four processes that can scatter $k_1, -k_1$ to $k_2, -k_2$.

How does all this help. Suppose $|k_1, -k_1\rangle$ and $|k_2, -k_2\rangle$ are only two states around. Then, a state like

$$\phi = \frac{|k_1, -k_1\rangle + |k_2, -k_2\rangle}{\sqrt{2}} \tag{8}$$

has energy $2\epsilon + \Delta_b$. That has lower energy than the individual states in the superposition. Δ_b is the binding energy. Now, let us remember what is Ω. It comes from electron–phonon interaction. Let us pause, develop this a bit in the next section and come back to our discussion.

3. Electron-phonon collisions

Recall we are interested in studying how a BCS molecule scatters phonons. For this we first understand how a normal electron scatters of a thermal phonon. We also derive electron–phonon interaction (*Fröhlich*) Hamiltonian and show how to calculate Ω in the above. The following section is a bit lengthy as it develops ways to visualize how a thermal phonon scatters an electron.

Consider phonons in a crystalline solid. We first develop the concept of a phonon packet. To fix ideas, we start with the case of one-dimensional lattice potential. Consider a periodic potential with period a:

$$U(x) = \sum_{l=1}^{n} V(x - a_l) = \sum V(x - la),$$

where

$$V(x) = V_0 \cos^2\left(\frac{\pi x}{a}\right), \quad \frac{-a}{2} \leq x \leq \frac{a}{2} \tag{9}$$

$$= 0 \quad |x| \geq \frac{a}{2}. \tag{10}$$

The potential is shown in **Figure** 7.

Now, consider how potential changes when we perturb the lattice sites from their equilibrium position, due to lattice vibrations:

$$\Delta U(x) = \sum V'(x - a_l)\Delta a_l.$$

For a phonon mode with wavenumber k

$$\Delta a_l = A_k \frac{1}{\sqrt{n}} \exp(ika_l), \tag{11}$$

we have

$$\Delta U(x) = A_k \exp(ikx) \frac{1}{\sqrt{n}} \underbrace{\sum V'(x - a_l) \exp(-ik(x - a_l))}_{p(x)},$$

$$= A_k \exp(ikx) \frac{1}{\sqrt{n}} \underbrace{\sum V'(x - la) \exp(-ik(x - la))}_{p(x)},$$

where $p(x)$ is the periodic with period a. Note

$$V'(x) = -V_0 \frac{2\pi}{a} \sin\left(\frac{2\pi x}{a}\right), \frac{-a}{2} \leq x \leq \frac{a}{2}$$

$$= 0|x| \geq \frac{a}{2}.$$

Using Fourier series, we can write

$$p(x) = a_0 + \sum_r a_r \exp\left(i\frac{2\pi r x}{a}\right).$$

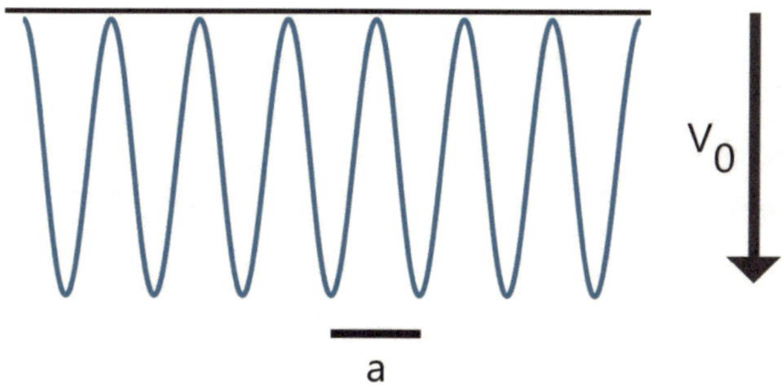

Figure 7.
Depiction of the periodic potential in Eq. (1).

We can determine a_0 by $a_0 = \frac{1}{a}\int_{-\frac{a}{2}}^{\frac{a}{2}} p(x)dx$, giving

$$a_0 = i\frac{V_0}{a}\int_{-\frac{a}{2}}^{\frac{a}{2}} \frac{2\pi}{a} \quad \sin\left(\frac{2\pi x}{a}\right) \sin(kx),$$

where $k = \frac{2\pi m}{na}$. This gives

$$a_0 = i\frac{2V_0}{a}\frac{1}{1-\left(\frac{m}{n}\right)^2} \sin\frac{m\pi}{n}.$$

We do not worry much about a_r for $r \neq 0$ as these excite an electron to a different band and are truncated by the band-gap energy. Now, note that, using equipartition of energy, there is $k_B T$ energy per phonon mode, giving

$$A_k = \sqrt{\frac{k_B T}{m}\frac{1}{\omega_k}} = \sqrt{\frac{k_B T}{m}\frac{1}{\omega_d \sin\left(\frac{\pi m}{n}\right)}}, \tag{12}$$

where ω_d is the Debye frequency.
Then, we get

$$\Delta U(x) = \frac{i}{\sqrt{n}}\underbrace{\left(\frac{2V_0}{a}\sqrt{\frac{k_B T}{m}\frac{1}{\omega_d}}\right)}_{\overline{V}_0} \exp(ikx). \tag{13}$$

At temperature of T = 3 K and $\omega_d = 10^{13}$ rad/s, we have.
$\sqrt{\frac{k_B T}{m}\frac{1}{\omega_d}} \sim .03\,^\circ$A; with a = 3 Å, we have

$$\Delta U \sim \frac{i}{\sqrt{n}}.01 V_0 \exp(ikx);$$

and with $V_0 = 10V$, we have

$$\Delta U \sim \frac{.1i}{\sqrt{n}} \exp(ikx)\, V.$$

We considered one phonon mode. Now, consider a phonon wave packet (which can also be thought of as a mode, localized in space) which takes the form

$$\Delta a_l = \frac{1}{n}\sum_k A_k \exp(ika_l),$$

where $k = m\Delta$ and $\Delta = \frac{2\pi}{na}$ and A_k as in Eq. (12). Then, the resulting deformation potential from Eq. (13) by summing over all phonon modes that build a packet becomes

$$\Delta U \sim \overline{V}_0 \frac{\sin^2\left(\frac{\pi x}{2a}\right)}{\left(\frac{\pi x}{2a}\right)}. \tag{14}$$

This deformation potential due to phonon wave packet is shown in **Figure 8**. The maximum value of the potential is around $\frac{\overline{V}_0}{\sqrt{2}}$.

3.1 Time dynamics and collisions

Of course phonons have a time dynamics given by their dispersion relation:

$$\Delta a_l(t) = \frac{1}{n}\sum_{k>0}\left(A_k \exp\left(ika_l\right)\exp\left(-i\omega_k t\right) + h.c\right). \tag{15}$$

With the phonon dispersion relation $\omega_k \sim vk$, where v is the velocity of sound, we get

$$\Delta U \sim \overline{V}_0 \frac{\sin^2\left(\frac{\pi(x-vt)}{2a}\right)}{\left(\frac{\pi(x-vt)}{2a}\right)}. \tag{16}$$

The deformation potential travels with velocity of sound and collides with an incoming electron. To understand this collision, consider a phonon packet as in Eq. (14) centred at the origin. The packet is like a potential hill. A electron comes along say at velocity v_g. If the velocity is high enough (kinetic energy $\frac{1}{2}mv_g^2 > \frac{e\overline{V}_0}{\sqrt{2}}$) to climb the hill, it will go past the phonon as in (b) in **Figure 9**, or else it will slide back, rebound of the hill and go back at the same velocity v_g as in (a) in **Figure 9**.

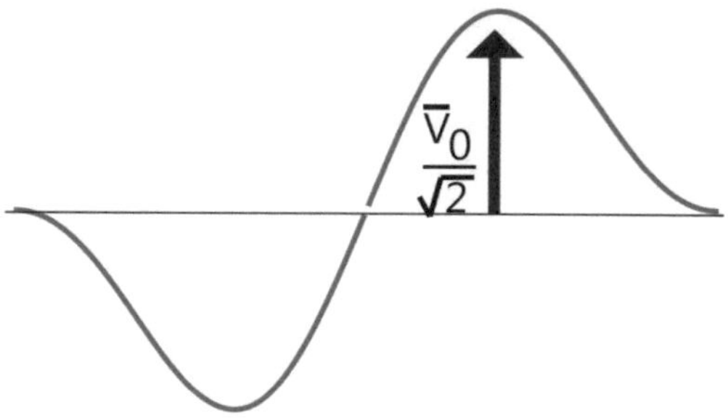

Figure 8.
Depiction of the deformation potential in Eq. (14).

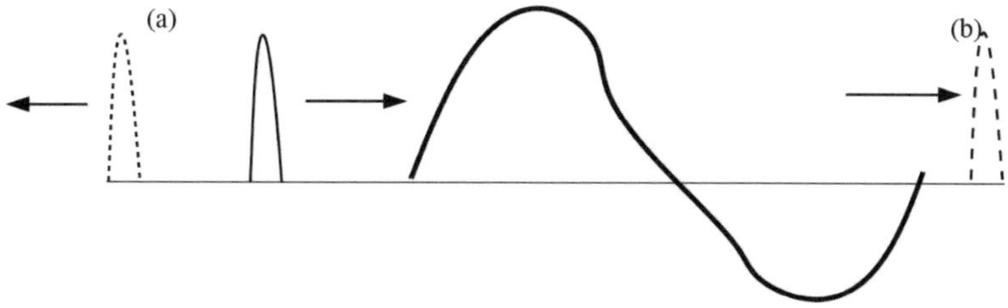

Figure 9.
Depiction on how an incoming electron goes past the deformation potential (b); if its velocity is sufficient, else it slides back and rebounds as in (a).

In the above, we assumed phonon packet is stationary; however, it moves with velocity v. Now, consider two scenarios. In the first one, the electron and phonon are moving in opposite direction and collide. This is shown in **Figure 10**.

In the phonon frame the electron travels towards it with velocity $v_g + v$. If the velocity is high enough (kinetic energy $\frac{1}{2}m(v_g + v)^2 > \frac{e\overline{V}_0}{\sqrt{2}}$) to climb the hill, it will go past the phonon with velocity $v_g + v$ (the resulting electron velocity in lab frame is just v_g). Otherwise, it slides back, rebounds and goes back with velocity $v_g + v$ (the velocity in lab frame is $v_g + 2v$). Therefore, electron has gained energy:

$$\Delta E = m v_g v \tag{17}$$

and by conservation of energy, the phonon has lost energy, lowering its temperature.

In the second case, electron and phonon are traveling in the same direction. This is shown in **Figure 11**. In the frame of phonon, electron travels towards the phonon with velocity $v_g - v$. If the velocity is high enough (kinetic energy $\frac{1}{2}m(v_g - v)^2 > \frac{e\overline{V}_0}{\sqrt{2}}$) to climb the hill, it will go past the phonon with velocity $v_g - v$. The velocity in lab frame is v_g. Otherwise, it slides back, rebounds and goes back with velocity $v_g - v$. Then, the velocity in lab frame is $v_g - 2v$. Therefore, electron has lost energy, and by conservation of energy, the phonon has gained energy, raising its temperature.

Thus, we have shown that electron and phonon can exchange energy due to collisions. Now, everything is true as in statistical mechanics, and we can go on to derive *Fermi-Dirac* distribution for the electrons [4–7].

All our analysis has been in one dimension. In two or three dimensions, the phonon packets are phonon tides (as in ocean tides). Let us fix ideas with two dimensions, and three dimensions follow directly. Consider a two-dimensional periodic potential with period a:

$$U(x,y) = \sum_{lm} V(x - a_l, y - a_m) = \sum_{lm} V(x - la, y - ma). \tag{18}$$

$$\begin{aligned} V(x,y) &= V_0 \cos^2\left(\frac{\pi x}{a}\right)\cos^2\left(\frac{\pi y}{a}\right), \quad \frac{-a}{2} \le x, y \le \frac{a}{2} \\ &= 0 \quad |x|, |y| \ge \frac{a}{2}. \end{aligned} \tag{19}$$

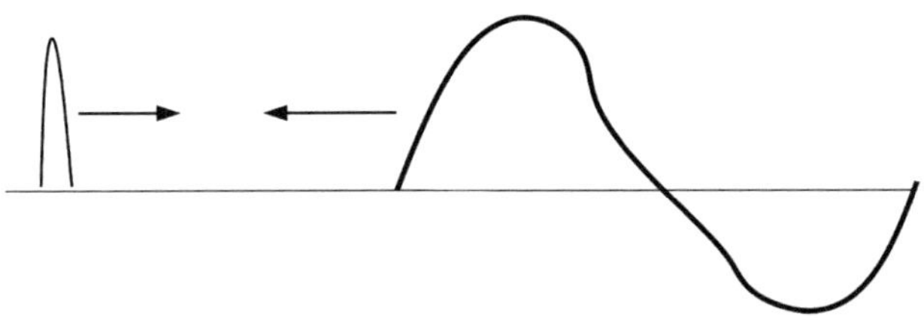

Figure 10.
Depiction on how an electron and phonon traveling towards each other collide.

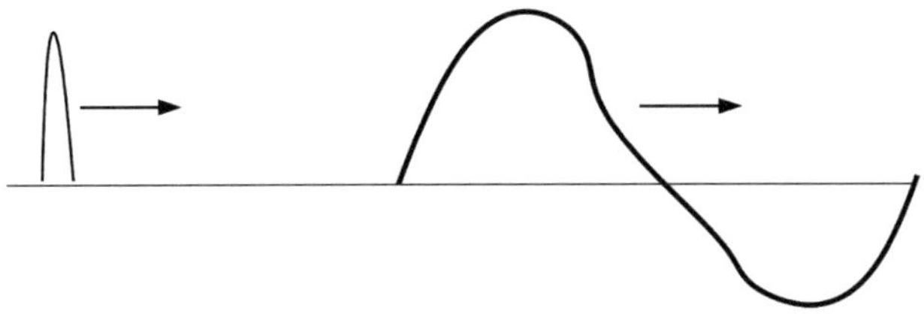

Figure 11.
Depiction on how an electron and phonon traveling in same direction collide.

Now, consider how potential changes when we perturb the lattice sites from their equilibrium position, due to lattice vibrations:

$$\Delta U(x,y) = \sum_{lm} V_x(x - a_l, y - a_m)\Delta a_l + V_y(x - a_l, y - a_m)\Delta a_m.$$

Let us consider phonons propagating along x direction. Then, Δa_l constitutes longitudinal phonons, while Δa_m constitutes transverse phonons. Transverse phonons do not contribute to deformation potential as can be seen in the following. Let us focus on the transverse phonons. Then,

$$\Delta a_m = A_k \frac{1}{\sqrt{n}} \exp\left(ik_x a_l\right). \tag{20}$$

We have due to Δa_m

$$\Delta U(x,y) = A_k \exp\left(ik_x x\right) \frac{1}{\sqrt{n}} \underbrace{\sum V_y(x - a_l, y - a_m) \exp\left(-ik(x - a_l)\right)}_{p(x,y)},$$

where $p(x,y)$ is the periodic with period a.
Note

$$V_y(x,y) = -V_0 \frac{2\pi}{a} \sin\left(\frac{2\pi y}{a}\right) \cos\left(\frac{\pi x}{a}\right)^2, \ \frac{-a}{2} \le x, y \le \frac{a}{2}$$
$$= 0 \quad |x|, |y| \ge \frac{a}{2}.$$

Using Fourier series, we can write

$$p(x,y) = a_0 + \sum_{r,s} a_{rs} \exp\left(i\left(\frac{2\pi r x}{a} + \frac{2\pi s y}{a}\right)\right).$$

We can determine a_0 by $a_0 = \frac{1}{a^2}\int_{-\frac{a}{2}}^{\frac{a}{2}}\int_{-\frac{a}{2}}^{\frac{a}{2}} p(x,y)dxdy$, giving $a_0 = 0$. Hence, transverse phonons do not contribute. The contribution of longitudinal phonons is same as in 1D case. As before consider a wave packet of longitudinal phonons propagating along x direction:

$$\Delta a_l = \frac{1}{n}\sum_k A_k \exp\left(ika_l\right),$$

which gives us a deformation potential as before:

$$\Delta U(x,y) \sim \overline{V}_0 \frac{\sin^2\left(\frac{\pi(x - vt)}{2a}\right)}{\left(\frac{\pi(x - vt)}{2a}\right)}, \tag{21}$$

which is same along y direction and travels with velocity v along the x direction except now the potential is like a tide in an ocean, as shown in **Figure 12**.

Since deformation potential is a tide, electron–phonon collisions do not have to be head on; they can happen at oblique angles, as shown in **Figure 13** in a top view (looking down). The velocity of electron parallel to tide remains unchanged, while velocity perpendicular to tide gets reflected. If the perpendicular velocity is large enough, the electron can jump over the tide and continue as shown by a dotted line in **Figure 13**. Imagining the tide in three dimensions is straightforward. In three

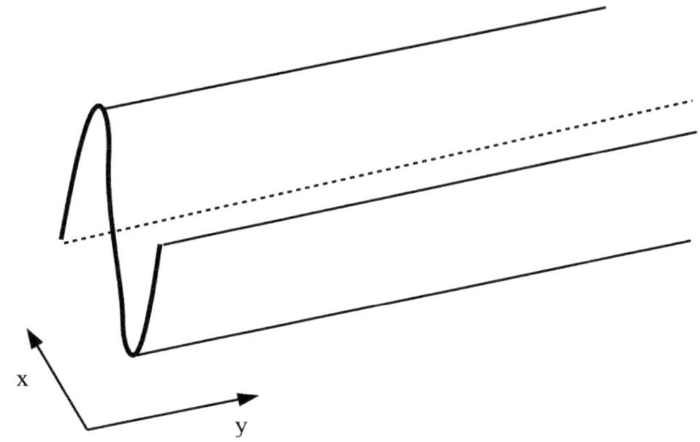

Figure 12.
Depiction of the deformation potential tide as shown in Eq. (21).

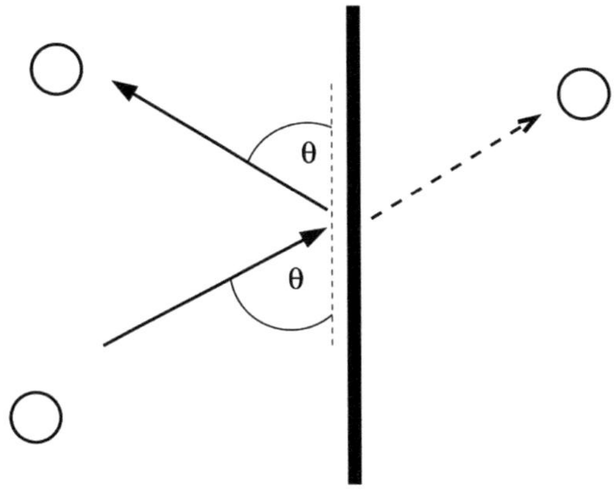

Figure 13.
The top view collision of an electron with a deformation potential tide at an angle θ.

dimensions, the deformation potential takes the form a wind gust moving in say x direction.

We described how a normal electron scatters phonons. Now, let us go back to our discussion on electron-phonon interaction and recall a phonon $\exp(ikx)$ which produces a deformation potential $\Delta U(x)$ as in Eq. 13. Now in three dimensions, it is

$$\Delta U(x) = \frac{i}{\sqrt{n^3}} \frac{V_0}{a} A_k(\exp(ikx)\exp(-i\omega_k t) - h.c), \tag{22}$$

where n^3 is the number of lattice points.

Using $\frac{1}{2}M\omega_k^2 A_k^2 = n_k \hbar\omega_k$ (there are n_k quanta in the phonon), where M is the mass of ion, ω_k phonon frequency and replacing A_k, we get

$$\Delta U(x) = \underbrace{\frac{i}{\sqrt{n^3}} \frac{V_0}{a} \sqrt{\frac{2\hbar}{M\omega_k}}}_{c} (\sqrt{n_k}\exp(ikx)\exp(-i\omega_k t) - h.c). \tag{23}$$

Thus, electron-phonon coupling Hamiltonian is of form

$$\underbrace{\frac{c}{\sqrt{n^3}}}_{\Omega} i\left(b \exp\left(ikx\right) - b^\dagger \exp\left(-ikx\right)\right), \tag{24}$$

where $b, b^\dagger$ are the annihilation and creation operators for phonon.

Using a cosine potential with $V_0 \sim 10$ V, we can approximate Coulomb potential. Then, with $a \sim 3A^\circ$ and $M \sim 20$ proton masses, we have $c \sim 1$ V.

We just derived an expression for the electron-phonon interaction (*Fröhlich*) Hamiltonian in Eq. (24) and showed how to calculate the constant c.

We said there are only two states, $|k_1, -k_1\rangle$ and $|k_2, -k_2\rangle$. In general we have for $i = 1, \ldots, \mathcal{N}$, $|k_i, -k_i\rangle$ states on the Fermi sphere as shown in **Figure 14A**, and if we form the state

$$\phi = \frac{1}{\sqrt{N}} \sum_i |k_i, -k_i\rangle, \tag{25}$$

it has energy $2\epsilon + (\mathcal{N} - 1)\Delta_b$.

The states do not have to be exactly on a Fermi surface as shown in **Figure 14A**; rather, they can be in an annulus around the Fermi sphere as shown in **Figure 14B**. When $|k_1, -k_1\rangle$ and $|k_2, -k_2\rangle$ are not both on the Fermi surface (rather in an annulus) such that the energy of $|k_1, -k_1\rangle$ is $E_1 = 2\epsilon_1$ and the energy of $E_3 = |k_2, -k_2\rangle$ is $2\epsilon_2$, with $\varepsilon_1 \neq \varepsilon_2$, then the formula of scattering amplitude (as shown in detail below) is modified to $\Delta_b = \frac{4\epsilon_d\Omega^2}{\Delta\epsilon^2 - \epsilon_d^2}$, where $\epsilon_1 - \epsilon_2 = \Delta\epsilon = \hbar\Delta\omega$. As long as $\Delta\omega < \omega_d$, in BCS theory, we approximate $\Delta_b \sim -\frac{4\Omega^2}{\epsilon_d}$. Therefore, if we take an annulus in **Figure 14B** to be of width ω_d, we get a total number of states in the annulus $\mathcal{N}$ to be $\frac{\mathcal{N}}{n^3} \sim \frac{\omega_d}{\omega_F}$, where n^3 is the total number of k points in the Fermi sphere and $\epsilon_F = \hbar\omega_F$ is the Fermi energy. This gives a binding energy $\Delta_b = \sim \frac{c^2}{\epsilon_F}$. With the Fermi energy $\epsilon_F \sim 10$ eV, the binding energy is ~meV. Thus, we have shown how phonon-mediated interaction helps us bind an electron pair with energy ~meV.

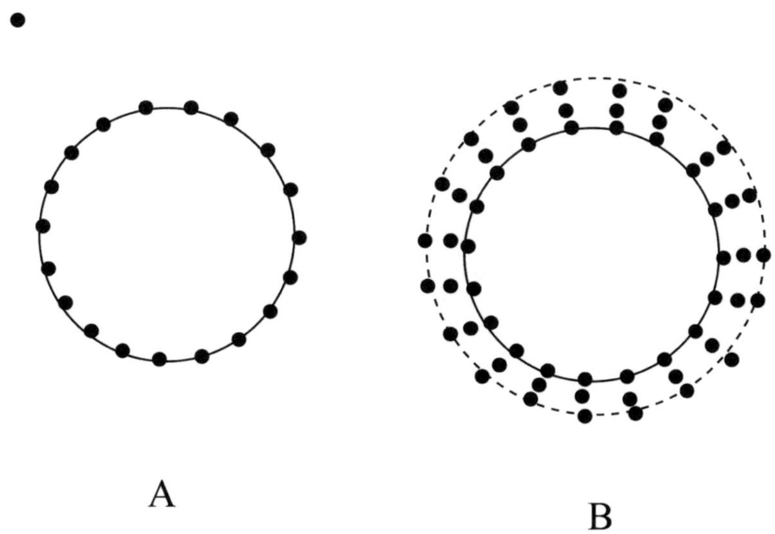

Figure 14.
(A) Electron pairs on the fermi surface and (B) electron pairs in an annulus around the fermi surface.

This paired electron state is called Cooper pair. Now, the plan is we bind many electrons and make a big molecule called BCS ground state.

But before we proceed, a note of caution is in order when we use the formula $\Delta_b = \frac{4\epsilon_d \Omega^2}{\Delta\epsilon^2 - \epsilon_d^2}$. For this we return to phonon scattering of $|k_1, -k_1\rangle$ and $|k_2, -k_2\rangle$. Consider when $E_1 \neq E_3$. Observe $E_1 = 2\epsilon_1 = 2\hbar\omega_1$, $E_3 = 2\epsilon_2 = 2\hbar\omega_2$, $E_2 = \epsilon_1 + \epsilon_2$ and $\hbar\omega_d$ is the energy of the emitted phonon. All energies are with respect to Fermi surface energy ϵ_F. The state of the three-level system evolves according to the Schrödinger equation:

$$\dot{\psi} = -i \begin{bmatrix} 2\omega_1 & d\exp(-i\omega_d t) & 0 \\ d\exp(i\omega_d t) & \omega_1 + \omega_2 & d\exp(i\omega_d t) \\ 0 & d\exp(-i\omega_d) & 2\omega_2 \end{bmatrix} \psi. \tag{26}$$

We proceed into the interaction frame of the natural Hamiltonian (system energies) by transformation:

$$\phi = \exp\left(it \begin{bmatrix} 2\omega_1 & 0 & 0 \\ 0 & \omega_1 + \omega_2 & 0 \\ 0 & 0 & 2\omega_2 \end{bmatrix} \right) \psi. \tag{27}$$

This gives for $\Delta E_1 = \omega_d - \underbrace{(\omega_1 - \omega_2)}_{\Delta\omega}$ and $\Delta E_2 = \omega_d + \Delta\omega$

$$\dot{\phi} = \underbrace{-i \begin{bmatrix} 0 & \exp(-i\Delta E_1\ t)d & 0 \\ \exp(i\Delta E_1\ t)d & 0 & \exp(i\Delta E_2\ t)d \\ 0 & \exp(-i\Delta E_2\ t)d & 0 \end{bmatrix}}_{H(t)} \phi. \tag{28}$$

We evaluate the effective evolution of $H(t)$ in period $\Delta t = \frac{2\pi}{\omega_d}$. After Δt, the system evolution is

$$\phi(\Delta t) = \exp\left(\underbrace{\int_0^{\Delta t} H(\sigma)d\sigma + \frac{1}{2}\int_0^{\Delta t} \left[H(\sigma_1), \int_0^{\sigma_1} H(\sigma_2)d\sigma_2 \right] d\sigma_1}_{\overline{H}\ \Delta t} \right) \phi(0). \tag{29}$$

Let us calculate $\overline{H}_{12}$. Assuming $|\Delta\omega| \ll \omega_d$, then

$$\overline{H}_{12} = \frac{-i}{\Delta t}\int_0^{\Delta t} \exp(-i\Delta E_1\ t)d = i\frac{\Delta\omega}{\omega_d}d. \tag{30}$$

Similarly

$$\overline{H}_{23} = \frac{-i}{\Delta t}\int_0^{\Delta t} \exp(-i\Delta E_2\ t)d = -i\frac{\Delta\omega}{\omega_d}d, \tag{31}$$

and finally

$$\begin{aligned}\overline{H}_{13} &= \frac{-d^2}{2\Delta t}\left\{\int_0^{\Delta t} \exp\left(-i\Delta E_1\ t\right)\int_0^{\tau} \exp\left(i\Delta E_2\ t\right) - \int_0^{\Delta t} \exp\left(i\Delta E_2\ t\right)\int_0^{\tau} \exp\left(-i\Delta E_1\ t\right)\right\} \\ &= \frac{id^2}{2\Delta t}\left(\frac{1}{\omega_d+\Delta\omega}+\frac{1}{\omega_d-\Delta\omega}\right)\int_0^{\Delta t} \exp\left(i2\Delta\omega\ t\right) = -i\frac{-d^2\omega_d\ \exp\left(i\Delta\omega\Delta t\right)}{\omega_d^2-\Delta\omega^2} \\ &\sim -i\frac{-d^2\exp\left(i\Delta\omega\Delta t\right)}{\omega_d}.\end{aligned}$$

Then, from Eq. (27), we get

$$\psi(\Delta T) = \exp\left(-i\Delta\begin{bmatrix}2\omega_1 & 0 & 0 \\ 0 & \omega_1+\omega_2 & 0 \\ 0 & 0 & 2\omega_2\end{bmatrix}\right)\ \exp\left(\overline{H}\Delta t\right)\ \psi(0) = \exp\left(H^{eff}\Delta t\right)\psi(0), \tag{32}$$

where

$$H^{eff} = -i\begin{bmatrix}2\omega_1 & -\frac{\Delta\omega}{\omega_d}d\exp\left(\frac{i}{2}\Delta\omega\Delta t\right) & \frac{-d^2}{\omega_d} \\ -\frac{\Delta\omega}{\omega_d}d\exp\left(-\frac{i}{2}\Delta\omega\Delta t\right) & \omega_1+\omega_2 & \frac{\Delta\omega}{\omega_d}d\exp\left(-\frac{i}{2}\Delta\omega\Delta t\right) \\ \frac{-d^2}{\omega_d} & \frac{\Delta\omega}{\omega_d}d\exp\left(\frac{i}{2}\Delta\omega\Delta t\right) & 2\omega_2\end{bmatrix}. \tag{33}$$

Observe in the above the term H_{13}^{eff} gives us the attractive potential responsible for superconductivity [3]. We say the electron pair $k_1,\ -k_1$ scatters to $k_2,\ -k_2$ at rate $-\frac{d^2}{\omega_d}$. But observe $d \ll \Delta\omega$, that is, term H_{13}^{eff} is much smaller than terms H_{12}^{eff} and H_{23}^{eff}: then, how are we justified in neglecting these terms. This suggests our calculation of scattering into an annulus around the Fermi surface requires caution and our expression for the binding energy may be high as binding deteriorates in the presence of offset. However, we show everything works as expected if we move to a wave-packet picture. The key idea is what we call *offset averaging*, which we develop now.

We have been talking about electron waves in this section. Earlier, we spent considerable time showing how electrons are wave packets confined to local potentials. We now look for phonon-mediated interaction between wave packets. A wave packet is built from many k-states (k-points). These states have slightly different energies (frequencies) which make the packet moves. We call these different frequencies *offsets* from the centre frequency. Denote $\exp\left(ik_0x\right)p(x)$ as a wave packet centred at momentum k_0. The key idea is that due to local potential, the wave packet shuttles back and forth and comes back to its original state. This means on average that the energy difference between its k-points averages and the whole packet just evolved with frequency $\omega(k_0)$. We may say the packet is *stationary* in the well, evolving as $\exp\left(ik_0x\right)p(x) \rightarrow \exp\left(-i\omega(k_0)t\right)\exp\left(ik_0x\right)p(x)$. **Figure 15** picturizes the offsets getting averaged by showing wave packets standing in the local potential. All we are saying is that now the whole packet has energy ϵ_0. So now, we can study how packet pair (at fermi surface) centred at $k_1,\ -k_1$ scatters to $k_2,\ -k_2$. This is shown in **Figure 16A**. The scattering is through a phonon packet with width localized to the local well. Let us say our electron packet width is Debye frequency ω_d. If there are N k-points in a packet, the original scattering rate Δ_b gets modified

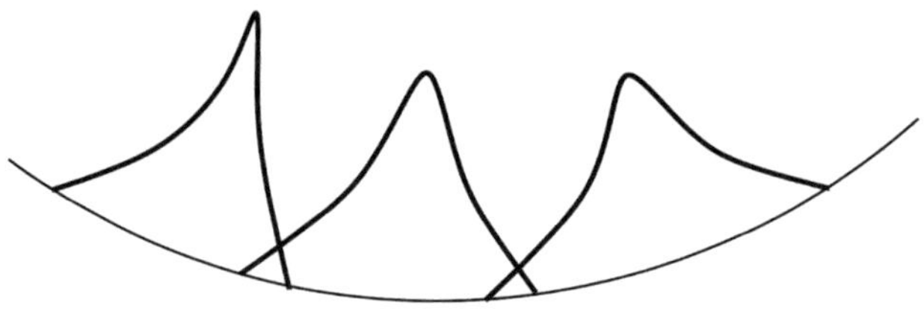

Figure 15.
The offsets getting averaged by showing wave packets standing in the local potential, bound by phonon-mediated interaction. Offset average so effectively packets are standstill; they do not move.

to $N\Delta_b$. If there are p packet pairs at the Fermi surface as shown in **Figure 16A**, then the state formed from superposition of packet pairs

$$\phi = \frac{1}{\sqrt{p}} \sum_i |k_i, \; -k_i\rangle \tag{34}$$

has binding energy $pN\Delta_b$. What is Np though is really just a $\mathcal{N}$ number of points in the annulus around the Fermi surface which we counted earlier. Hence, we recover the binding energy we derived between electrons earlier, but now it is binding of packets. Really, this way we do not have to worry about offsets when we bind; in packet land they average in local potential.

The wave packet in a potential well shuttles back and forth, which averages the offsets $\omega_i - \omega_j$ to zero. Then, the question of interest is how fast do we average these offsets compared to packet width which we take as Debye frequency ω_d. For s orbitals or waves, the bandwidth is ~10 eV giving Fermi velocity of 10^5 m/s, which for a characteristic length of the potential well as ~300 Å, corresponds to a packet shuttling time of around $\sim 10^{-13}$s, which is same as the Debye frequency ω_d, so we may say we average the offsets in a packet. By the time offsets have evolved significantly, the packet already returns to its original position, and we may say there are no offsets.

We saw how two electrons bind to form a Cooper pair. However, for a big molecule, we need to bind many electrons. How this works will be discussed now. The basic idea is with many electrons; we need space for electron wave packets to scatter to. For example, when there was only one packet pair at the Fermi surface, it could scatter into all possible other packet pairs, and we saw how we could then form a superposition of these states. Now, suppose we have $2p$ packet pairs possible on the Fermi surface. We begin with assuming p of these are occupied with

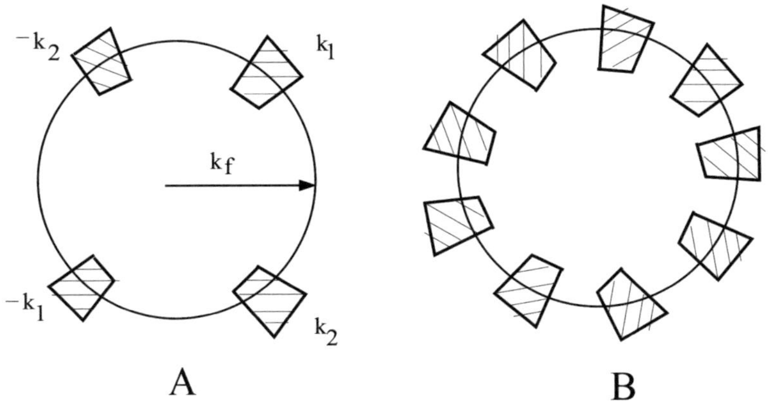

Figure 16.
(A) Packet pairs k_1, $-k_1$ and k_2, $-k_2$ on the Fermi surface and (B) many such pairs.

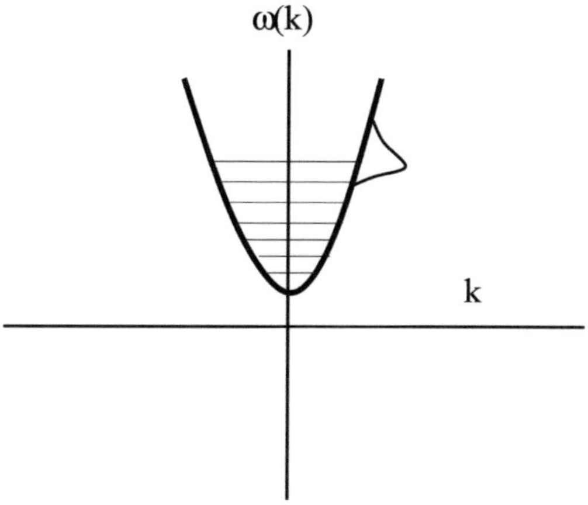

Figure 17.
Packets at the Fermi surface that has twice the number of k-states as electrons. This means half the packet sites will be empty.

electrons and remaining p empty. This way we create space for the p pairs to scatter into; otherwise, if all are full, how will we scatter? How do these empty spaces come about? We just form packets with twice the bandwidth as there are electrons. Then, half of these packets are empty as shown in **Figure 17**.

4. BCS ground state

Let k_F denote wavevector radius of the Fermi sphere. We describe electron wave packets formed from k-points (wavevectors) near the Fermi sphere surface. **Figure 18** shows the Fermi sphere with surface as thick circle and an annulus of thickness ω_d (energy units) shown in dotted lines. Pockets have **n**, k-points in radial direction ($\frac{\mathbf{n}}{2}$ inside and same outside the Fermi sphere) and $\mathbf{m}^2$ points in the tangential direction. **Figure 18** shows such a pocket enlarged with k-points shown in

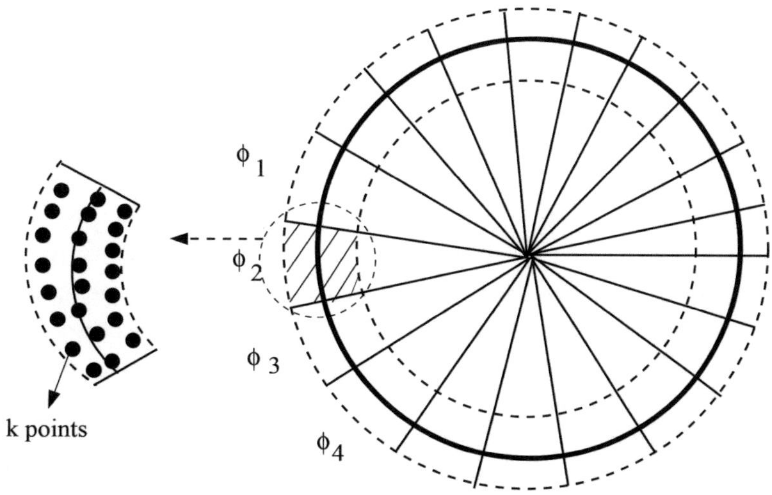

Figure 18.
Depiction of the Fermi sphere with surface as a thick circle and an annulus of thickness ω_d (energy units) shown in dotted lines. Pockets with k-points are shown in black dots. Superposition of these k-points in a pocket forms a wave packet.

black dots. We assume there are $4p$ such pockets with $N = \mathbf{nm}^2$ points in each pocket. In pocket s, we form the function

$$\phi_s(r) = \frac{1}{\sqrt{N}} \sum_j \exp\left(ik_j \cdot r\right) \tag{35}$$

ϕ_s has characteristic width a. From N points in a pocket, we can form N such functions by displacing ϕ_s to $\phi_s(r - ma)$, and putting these functions uniformly spaced over the whole lattice in their local potential wells. Thus, each pocket gives N wave packets orthogonal as they are nonoverlapping and placed uniformly over the lattice in their local potential wells. Furthermore, ϕ_s and $\phi_{s'}$ are orthogonal as they are formed from mutually exclusive k-points. The wave packet ϕ_s moves with a group velocity v, which is the Fermi velocity in a radial direction to the pocket from which it is formed. From wave packet ϕ_s and its antipodal packet ϕ_{-s}, we form the joint wave packet:

$$\Phi_s(r_1, r_2) = \phi_s(r_1)\phi_{-s}(r_2). \tag{36}$$

As we will see soon, Φ_s will be our Cooper pair.

As shown in last section, the Cooper pair Φ_s scatters to $\Phi_{s'}$ with rate $\mathcal{M} = -\frac{4d^2N}{\omega_d}$. Now, we study how to form BCS state with many wave packets. We present a counting argument. Observe ϕ_s is made of N k-points, and hence by displacement of ϕ_s, we have N nonoverlapping lattice sites (potential wells) where we can put copies of ϕ_s as described in Section 4. However of the N k-points, only $\frac{N}{2}$ are inside the Fermi sphere so we only have only $\frac{N}{2}$ wave packets. Therefore, $\frac{1}{2}$ of the wave-packet sites are empty. Hence, of the $2p$ possible Cooper pairs, only p are filled and p are not present. Hence, when Φ_s scatters to $\Phi_{s'}$, we have p choices for s'. Then, we can form a joint state of Cooper pairs present and write it as $\Phi_{i_1}\Phi_{i_2}...\Phi_{i_p}$. Doing a superposition of such states, we get the superconducting state:

$$\Psi = \sum \Phi_{i_1}\Phi_{i_2}...\Phi_{i_p}. \tag{37}$$

The binding energy of this state is $-\frac{4\hbar d^2 N p^2}{\omega_d}$ as each index in Ψ scatters to p states. Since $4pN$ are the total k-points in the annulus surrounding the Fermi surface. We have $\Omega = \frac{c}{\sqrt{n^3}}$ where c is of order 1 eV and n^3 is the total number of lattice sites in the solid. Then, $\frac{4Np}{n^3} \sim \frac{\omega_d}{\omega_F}$. Thus, the binding energy is $-pc\frac{c}{\hbar\omega_F}$ ($c\frac{c}{2\hbar\omega_F}$ per wave packet/electron). The Fermi energy $\hbar\omega_F$ is of order 10 eV, while $\hbar\omega_d \sim .1$ eV. Thus, binding energy per wave packet is of order .05 eV. The average kinetic energy of the wave packet is just the Fermi energy ϵ_F as it is made of superposition of $\frac{N}{2}$ points inside and $\frac{N}{2}$ points outside the Fermi sphere. If the wave packet was just formed from k-point inside the Fermi surface, its average energy would be $\epsilon_F - \frac{\hbar\omega_d}{4}$. Thus, we pay a price of $\frac{\hbar\omega_d}{4} \sim .025$ eV per wave packet; when we form our wave packet out of N k-points, half of which are outside the Fermi sphere. Thus, per electron wave packet, we have a binding energy of $\sim .025$ eV around 20 meV. Therefore, forming a superconducting state is only favorable if $c\frac{c}{\hbar\omega_F} > \frac{\hbar\omega_d}{4}$. Observe if c is too small, then forming the superconducting state is not useful, as the gain of binding energy is offsetted by the price we pay in having wave packets that have excursion outside the Fermi surface.

Next, we study how low-frequency thermal phonons try to break the BCS molecule. The electron wave packet collides with the phonon and gets deflected, which means the Cooper pair gets broken. Then, the superconducting state constitutes $p-1$ pairs and damaged pair. Each term in the state in Eq. (37) will scatter to $p(p-1)$ states, and the binding energy is $-\frac{4\hbar d^2 Np(p-1)}{\omega_d}$. The total binding energy has reduced by $\Delta = \frac{4\hbar d^2 Np}{\omega_d}$, the superconducting gap. The phonon with which the superconducting electron collides carries an energy $k_B T$. When this energy is less than Δ, the electron cannot be deflected and will not scatter as we cannot pay for the increase of energy. Therefore, superconducting electrons do not scatter phonons. However, this is not the whole story as the phonon can deflect the electron slightly from its course and over many such collisions break the Cooper pair; then, the energy budget Δ is paid in many increments of $k_B T$. These are small angle scattering events. Thus, there will be a finite probability q that a Cooper pair is broken, and we say we have excited a **Bogoliubov** [8]. This probability can be calculated using Boltzmann distribution. In a superconducting state with $2p$, wave packets on average $2pq$ will be damaged/deflected. This leaves only $p' = p(1-2q)$ good pairs which give an average energy per pair to be $(1-2q)\Delta$. The probability that the superconducting electron will be deflected of phonon is

$$q = \frac{\exp\left(-(1-2q)\Delta/k_B T\right)}{1 + \exp\left(-(1-2q)\Delta/k_B T\right)},$$

which gives

$$\ln\left(\frac{1}{q} - 1\right) = \frac{(1-2q)\Delta}{k_B T}. \tag{38}$$

When $k_B T_c = \Delta$, the above equation gives $q = \frac{1}{2}$ and gap $(1-2q)\Delta$ goes to zero. T_c is the critical temperature where superconducting transition sets in.

5. Molecular orbitals of BCS states

We now come to interaction of neighboring BCS molecules. In our picture of *local potentials*, we have electron wave packets in each potential well that are coupled as pairs by phonon to form the BCS ground state. What is important is that there is a BCS ground state in each potential well. When we bring two such wells in proximity, the ground state wave functions overlap, and we form a molecular orbital between these BCS orbitals. This is shown in **Figure 19**.

If Φ_1 and Φ_2 are the orbitals of Cooper pair in individual potential wells, then the overlap creates a transition $d = \langle \Phi_1 | 2eU | \Phi_2 \rangle$, where U is the potential well, with $2e$ coming from the electron pair charge. Then, the linear combination of atomic orbital (LCAO) $a\Phi_1 + b\Phi_2$ evolves as

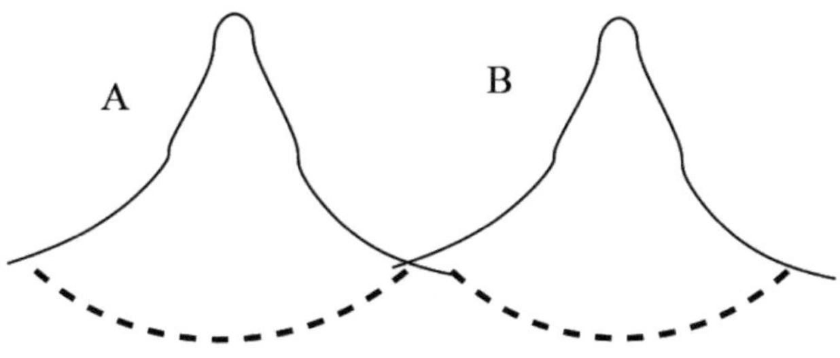

Figure 19.
Depiction of two BCS ground states in local potential wells separated in a weak link.

$$\hbar\frac{d}{dt}\begin{bmatrix} a \\ b \end{bmatrix} = -i\begin{bmatrix} \epsilon_1 - 2\Delta & d \\ d & \epsilon_2 - 2\Delta \end{bmatrix}\begin{bmatrix} a \\ b \end{bmatrix} \tag{39}$$

where ϵ_1, ϵ_2 is the Fermi energy of the electron pair in two orbitals. $\epsilon_1 = \epsilon_2$ unless we apply a voltage difference v between them; then, $\epsilon_1 - \epsilon_2 = v$. Let us start with $v = 0$ and $\begin{bmatrix} a \\ b \end{bmatrix} = \begin{bmatrix} 1 \\ 1 \end{bmatrix}$, and then nothing happens, but when there is phase difference $\begin{bmatrix} a \\ b \end{bmatrix} = \begin{bmatrix} 1 \\ \exp(i\phi) \end{bmatrix}$, then $\begin{bmatrix} a \\ b \end{bmatrix}$ evolves and we say we have supercurrent $I \propto \sin\phi$ between two superconductors. This constitutes the Josephson junction (in a Josephson junction, there is a thin insulator separating two BCS states or superconductors). Applying v generates a phase difference $\frac{d\phi}{dt} \propto v$ between the two orbitals which then evolves under d.

We talked about two BCS states separated by a thin insulator in a Josephson junction. In an actual superconductor, we have an array (lattice) of such localized BCS states as shown in **Figure 20**. Different phases ϕ_i induce a supercurrent as in Josephson junction.

If Φ_i denotes the local Cooper pairs, then their overlap creates a transition $d = \langle \Phi_i | 2eU | \Phi_{i+1} \rangle$, where U is the potential well, with $2e$ coming from the electron pair charge. Then, the LCAO $\sum a_i \Phi_i$ evolves as

$$\hbar\frac{d}{dt}\begin{bmatrix} a_1 \\ \vdots \\ a_n \end{bmatrix} = -i\begin{bmatrix} \epsilon_1 - 2\Delta & d & 0 \\ d & \ddots & \vdots \\ & d & \epsilon_n - 2\Delta \end{bmatrix}\begin{bmatrix} a_1 \\ \vdots \\ a_n \end{bmatrix} \tag{40}$$

where $\epsilon_i = \epsilon_0$ is the Fermi energy of the electron pair in BCS state Φ_i. ϵ_i is same unless we apply a voltage difference v to the superconductor. Eq. 40 is the tight-binding approximation model for superconductor.

What we have now is a new lattice of potential wells as shown in **Figure 20** with spacing of $b \sim 300$ Å as compared to the original lattice of $a \sim 3$ Å, which means 100 times larger. We have an electron pair at each lattice site. The state

$$\begin{bmatrix} a_1 \\ \vdots \\ a_n \end{bmatrix} = \frac{1}{\sqrt{n}}\begin{bmatrix} 1 \\ \vdots \\ 1 \end{bmatrix} \tag{41}$$

is the ground state of new lattice (Eq. 40). A state like

$$\begin{bmatrix} a_1 \\ \vdots \\ a_n \end{bmatrix} = \frac{1}{\sqrt{n}}\begin{bmatrix} \exp(ikx_1) \\ \vdots \\ \exp(ikx_n) \end{bmatrix}$$

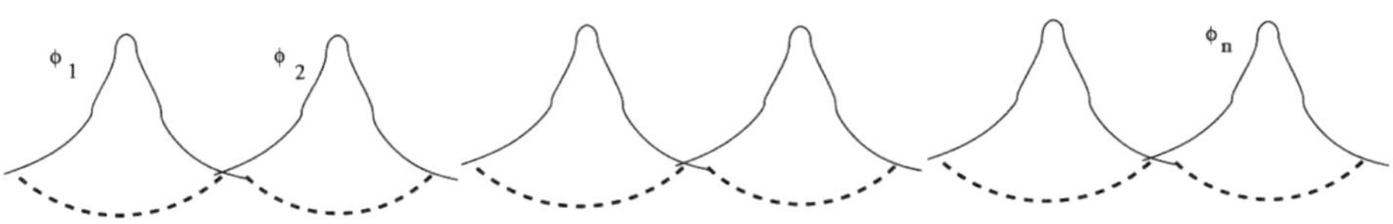

Figure 20.
Depiction of array of local BCS states in local potential wells with different phases.

has a momentum and constitutes the supercurrent. It has energy $\epsilon_k = \epsilon_0 - 2\Delta + 2d\cos kb$.

In the presence of electric field E, we get on the diagonal of RHS of Eq. 40, additional potential eEx_i. How does ϕ_k evolve under this field? Verify with $k(t) = k - \frac{eEt}{\hbar}$; $\phi_{k(t)}$ is a solution with eigenvalue $\epsilon_{k(t)}$. In general, in the presence of time varying $E(t)$, we have $k(t) = k - \frac{2e\int_0^t E(\tau)d\tau}{\hbar}$..

Now, consider the local BCS states in **Figure 20** put in a loop. If we turn on a magnetic field (say in time T) through the centre of the loop, it will establish a transient electric field in the loop given by

$$\int_0^T E(\tau) = \frac{Ba_r}{2\pi r},$$

where r is radius and a_r the area of the loop. Then, by the above argument, the wavenumber k of the BCS states is shifted by $\Delta k = \frac{(2e)\int_0^T E(\tau)}{\hbar} = \frac{(2e)Ba_r}{2\pi r\hbar}$. Since we have closed loop with $\Phi_0 = Ba_r$

$$\Delta k(2\pi r) = \frac{2e\Phi_0}{\hbar} = 2\pi, \tag{42}$$

giving

$$\Phi_0 = Ba_r = \frac{h}{2e}.$$

This is the magnetic quantum flux. When one deals with the superconducting loop or a hole in a bulk superconductor, it turns out that the magnetic flux threading such a hole/loop is quantized [9, 10] as just shown.

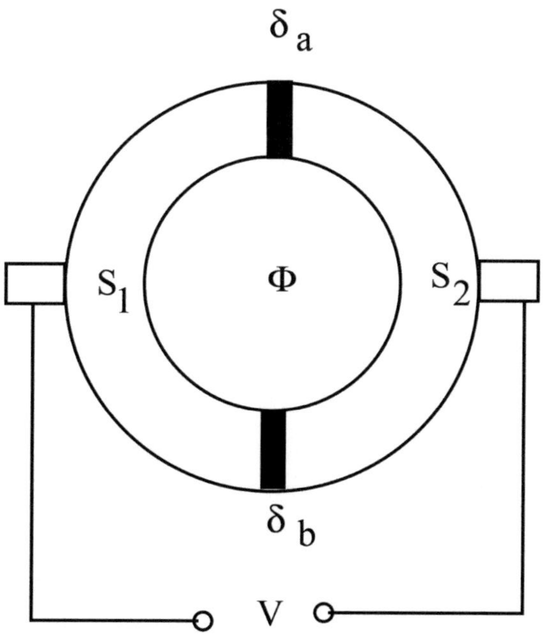

Figure 21.
Depiction of the schematic of a SQUID where two superconductors S_1 *and* S_2 *are separated by thin insulators.*

Figure 21 depicts the schematic of a superconducting quantum interference device (SQUID) where two superconductors S_1 and S_2 are separated by thin insulators. A small flux through the SQUID creates a phase difference in the two superconductors (see discussion on Δk above) leading to the flow of supercurrent. If an initial phase, δ_0 exists between the superconductors. Then, this phase difference after application of flux is from Eq. (42), $\delta_a = \delta_0 + \frac{e\Phi_0}{\hbar}$ across top insulator and $\delta_a = \delta_0 - \frac{e\Phi_0}{\hbar}$ across bottom insulator (see **Figure 21**). This leads to currents $J_a = I_0 \sin \delta_a$ and $J_b = I_0 \sin \delta_b$ through top and down insulators. The total current $J = J_a + J_b = 2I_0 \sin \delta_0 \cos \frac{e\Phi_0}{\hbar}$. This accumulates charge on one side of SQUID and leads to a potential difference between the two superconductors. Therefore, the flux is converted to a voltage difference. The voltage oscillates as the phase difference $\frac{e\Phi_0}{\hbar}$ goes in integral multiples of π for every flux quanta Φ_0. SQUID is the most sensitive magnetic flux sensor currently known. The SQUID can be seen as a flux to voltage converter, and it can generally be used to sense any quantity that can be transduced into a magnetic flux, such as electrical current, voltage, position, etc. The extreme sensitivity of the SQUID is utilized in many different fields of applications, including biomagnetism, materials science, metrology, astronomy and geophysics.

6. Meissner effect

When a superconductor placed in an magnetic field is cooled below its critical T_c, we find it expel all magnetic field from its inside. It does not like magnetic field in its interior. This is shown in **Figure 22**.

German physicists Walther Meissner and Robert Ochsenfeld discovered this phenomenon in 1933 by measuring the magnetic field distribution outside superconducting tin and lead samples. The samples, in the presence of an applied magnetic field, were cooled below their superconducting transition temperature, whereupon the samples canceled nearly all interior magnetic fields. A superconductor with little or no magnetic field within it is said to be in the Meissner state. The Meissner state breaks down when the applied magnetic field is too large. Superconductors can be divided into two classes according to how this breakdown occurs. In type-I superconductor if the magnetic field is above certain threshold H_c, no expulsion takes place. In type-II superconductors, raising the applied field past a critical value H_{c1} leads to a mixed state (also known as the vortex state) in which an increasing amount of magnetic flux penetrates the material, but there remains no

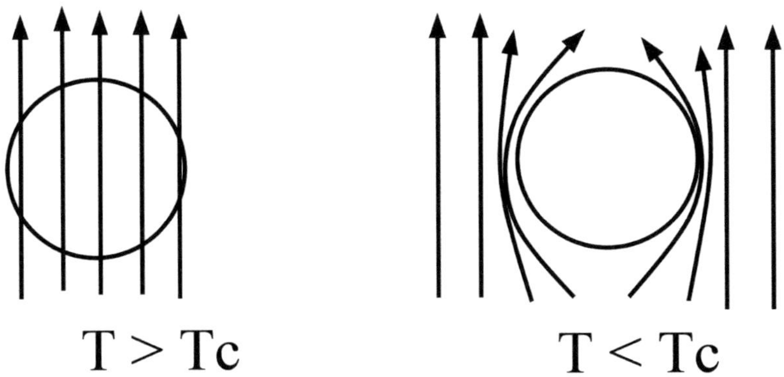

Figure 22.
Depiction of the Meissner effect whereby the magnetic field inside a superconductor is expulsed when we cool it below its superconducting temperature T_c.

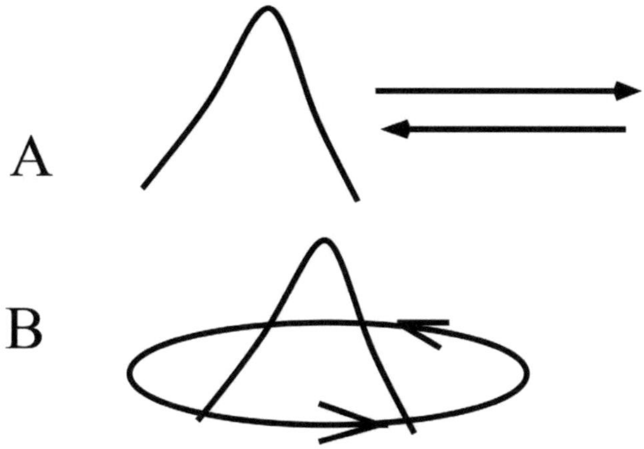

Figure 23.
(A) Depiction on how packets shuttle back and forth in local potential and (B) how they execute a cyclotron motion in the presence of a magnetic field.

resistance to the electric current as long as the current is not too large. At a second critical field strength H_{c2}, no magnetic field expulsion takes place. How can we explain Meissner effect?

We talked about how wave packets shuttle back and forth in local potentials and get bound by phonons to form a BCS molecule. In the presence of a magnetic field, they do not shuttle. Instead, they do cyclotron motion with frequency $\omega = \frac{eB}{m}$. This is shown in **Figure 23**. At a field of about 1 Tesla, this is about 10^{11} rad/s. Recall our packets had a width of $\omega_D \sim 10^{13}$ Hz and the shuttling time of packets was 10^{-13} s, so that the offsets in a packet do not evolve much in the time the packet is back. But, when we are doing cyclotron motion, it takes 10^{-11} s (at 1 T field) to come back, and by that time, the offsets evolve a lot, which means poor binding. It means in the presence of magnetic field we cannot bind well. Therefore, physics wants to get rid of magnetic field, bind and lower the energy. Magnetic field hurts binding and therefore it is expelled. But if the magnetic field is increased, then the cyclotron frequency increases, and at a critical value, our packet returns home much faster, allowing for little offset evolution; therefore, we can bind and there is no need to expulse the magnetic field. This explains the critical field.

7. Giaever tunnelling

When we bring two metals in proximity, separated by a thin-insulating barrier, apply a tiny voltage nd then the current will flow in the circuit. There is a thin-insulating barrier, but electrons will tunnel through the barrier. Now, what will happen if these metals are replaced by a superconductor? These are a set of experiments carried out by Norwegian-American physicist Ivar Giaever who shared the Nobel Prize in Physics in 1973 with Leo Esaki and Brian Josephson “for their discoveries regarding tunnelling phenomena in solids”. What he found was that if one of the metals is a superconductor, the electron cannot just come in, as there is an energy barrier of Δ, the superconducting gap. Your applied voltage has to be at least as big as Δ for tunneling to happen. This is depicted in **Figure 24**. Let us see why this is the case.

Recall in our discussion of superconducting state that we had $2p$ pockets of which p were empty. We had a binding energy of $-\frac{4\hbar d^2 N p^2}{\omega_d}$. What will it cost to

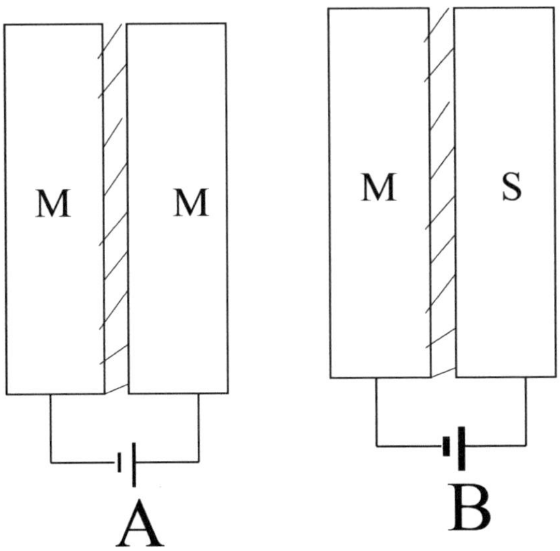

Figure 24.
(A) Depiction on how a tiny voltage between two metals separated by an insulating barrier generates current that goes through an insulating barrier through tunneling. (B) If one of the metals is a superconductor, then the applied voltage has to be at least as big as the superconducting gap.

Figure 25.
Depiction on how upon tunneling an extra electron, shown in dashed lines, enters the superconducting state.

bring in an extra electron as shown in **Figure 25**? It will go in one of the empty pockets, and then we have only $p-1$ pockets left to scatter to reducing the binding energy to $-\frac{4\hbar d^2 Np(p-1)}{\omega_d}$ with a change $\Delta = \frac{4\hbar d^2 Np}{\omega_d}$. Therefore, the new electron raises the energy by Δ, and therefore to offset this increase of energy, we have to apply a voltage as big as Δ for tunneling to happen.

8. High T_c in cuprates

High-temperature superconductors (abbreviated high-T_c or HTS) are materials that behave as superconductors at unusually high temperatures. The first high-T_c superconductor was discovered in 1986 by IBM researchers Georg Bednorz and K. Alex Müller, who were awarded the 1987 Nobel Prize in Physics "for their important break-through in the discovery of superconductivity in ceramic materials".

Whereas "ordinary" or metallic superconductors usually have transition temperatures (temperatures below which they are superconductive) below 30 K (243.2°C) and must be cooled using liquid helium in order to achieve superconductivity, HTS have been observed with transition temperatures as high as 138 K (135°C) and can be cooled to superconductivity using liquid nitrogen. Compounds of copper and oxygen (so-called cuprates) are known to have HTS properties, and the term high-temperature superconductor was used interchangeably with cuprate superconductor. Examples are compounds such as lanthanum strontium copper oxide (LASCO) and neodymium cerium copper oxide (NSCO).

Let us take lanthanum copper oxide La_2CuO_4, where lanthanum donates three electrons and copper donates two and oxygen accepts two electrons and all valence is satisfied. The Cu is in state Cu^{2+} with electrons in d^9 configuration [11]. d orbitals are all degenerated, but due to crystal field splitting, this degeneracy is broken, and $d_{x^2-y^2}$ orbital has the highest energy and gets only one electron (the ninth one) and forms a band of its own. This band is narrow, and hence all k states get filled by only one electron each and form Wannier packets that localize electrons on their respective sites. This way electron repulsion is minimized, and in the limit $U \gg t$ (repulsion term in much larger than hopping), we have Mott insulator and antiferromagnetic phase.

When we hole/electron dope, we remove/add electron to $d_{x^2-y^2}$ band. For example, $La_{2-x}Sr_xCuO_4$ is hole doped as Sr has valence 2 and its presence further removes electrons from copper. $Nd_{2-x}Ce_xCuO_4$ is electron doped as Ce has valence 4 and its presence adds electrons from copper. These extra holes/electrons form packets. When we discussed superconductivity, we discussed packets of width ω_d. In d-bands, a packet of this width has many more k states as d-bands are narrow and only 1–2 eV thick. This means N (k-points in a packet) is very large and we have much larger gap Δ and T_c. This is a way to understand high T_c, d-wave packets with huge bandwidths. This is as shown in **Figure 26**. As we increase doping and add

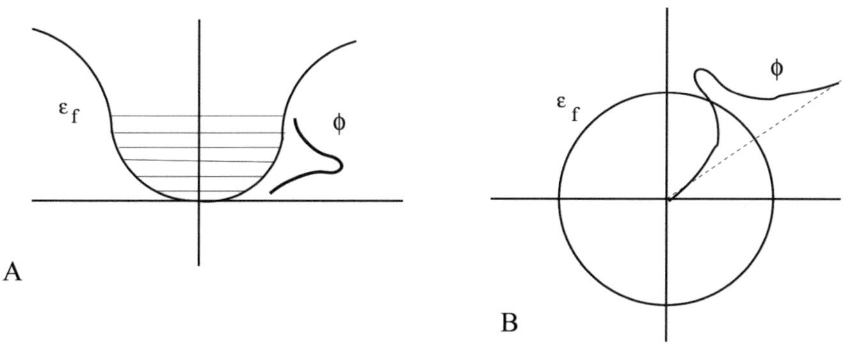

Figure 26.
(A) Depiction on how in narrow d *band the electron wave packet comprises all* k *points to minimize repulsion. (B) The wave packet in two dimensions with wave packet formed from superposition of k-points (along the k-direction) inside the Fermi sphere.*

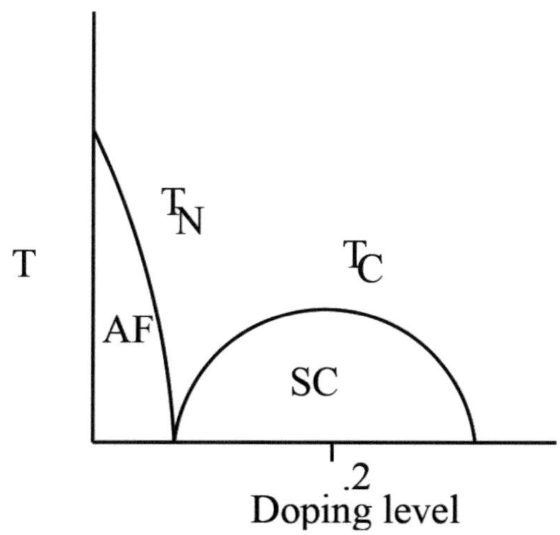

Figure 27.
The characteristic phase diagram of a high T_c *cuprate. Shown are the antiferromagnetic insulator (AF) phase on the left and* dome-shaped *superconducting phase (SC) in the centre.*

more electrons, the packet width further increases till it is $\gg\omega_d$; then, we have significant offset evolution, and binding is hurt leading to loss of superconductivity. This explains the *dome* characteristic of superconducting phase, whereby superconductivity increases and then decreases with doping. The superconducting *dome* is shown in **Figure 27**.

Author details

Navin Khaneja
IIT Bombay, Powai, India

*Address all correspondence to: navinkhaneja@gmail.com

References

[1] Tinkham M. Introduction to Superconductivity. 2nd ed. New York, USA: McGraw Hill; 1996

[2] De Gennes PG. Superconductivity of Metals and Alloys. New York, USA: W.A. Benjamin, Inc; 1966

[3] Bardeen J, Cooper L, Schriffer JR. Theory of superconductivity. Physical Review. 1957;**108**(5):1175

[4] Kittel C. Introduction to Solid State Physics. 8th ed. John Wiley and Sons; 2005

[5] Ashcroft NW, Mermin D. On a new method in the theory of superconductivity. Solid State Physics. Harcourt College Publishers; 1976

[6] Simon S. Oxford Solid State Basics. New Delhi, India: Oxford University Press; 2013

[7] Kittel C, Kroemer H. Thermal Physics. Orlando, USA: Freeman and Co; 2002

[8] Bogoliubov NN. Nuovo Cimento. 1958;7:794

[9] Deaver B, Fairbank W. Experimental evidence for quantized flux in superconducting cylinders. Physical Review Letters. 1961;7(2):43-46

[10] Doll R, Näbauer M. Experimental proof of magnetic flux quantization in a superconducting ring. Physical Review Letters. 1961;7(2):51-52

[11] Khomskii D. Transition Metal Compounds. Cambridge University Press; 2014